Y0-BDB-618

Fallout

Fallout

The Environmental Consequences
of the World Trade Center Collapse

Juan González

THE NEW PRESS
NEW YORK

© 2002 by Juan González
All rights reserved.
No part of this book may be reproduced, in any form,
without written permission from the publisher.

Published in the United States by The New Press, New York, 2002
Distributed by W. W. Norton & Company, Inc., New York

ISBN 1-56584-754-7 (hc.)
CIP data available

The New Press was established in 1990 as a not-for-profit alternative to the
large, commercial publishing houses currently dominating the book publishing industry. The New Press operates in the public interest rather than
for private gain, and is committed to publishing, in innovative ways, works
of educational, cultural, and community value that are often deemed in-
sufficiently profitable.

The New Press, 450 West 41st Street, 6th floor, New York, NY 10036
www.thenewpress.com

Printed in the United States of America

2   4   6   8   10   9   7   5   3   1

For Gabriela

Contents

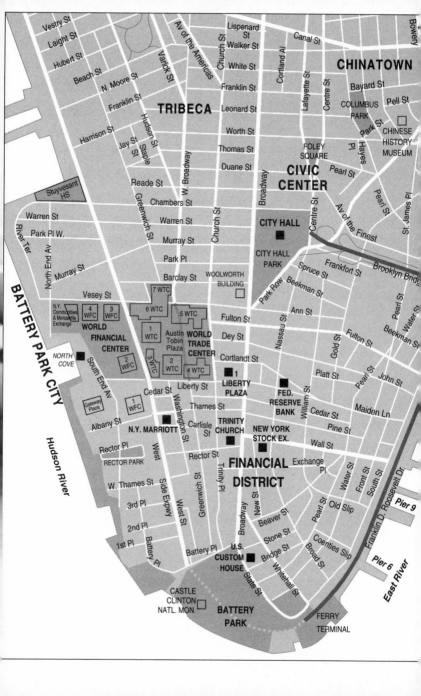

Introduction

On September 17, 2001, less than one week after the World Trade Center collapse, tens of thousands of office workers returned to their jobs near Ground Zero after receiving the go-ahead from federal and local safety officials.

At the time, federal and city government wanted New York and the rest of the nation, which had been virtually paralyzed in the days after the September 11 terrorist attacks, to return to normal as quickly as possible. President George W. Bush, New York Mayor Rudolph Giuliani, and other leaders needed to show the world that the United States would not be intimidated by terrorism. There was another, more pressing imperative at work, however: the longer that Wall Street and the nation's chief financial markets remained closed, the greater the likelihood of a stock meltdown and perhaps long-lasting damage to investors and the U.S. economy.

To achieve a rapid return to normalcy the government needed to persuade a jittery public that it was safe for civilians to reoccupy the scores of commercial skyscrapers and residential buildings in lower Manhattan. With half a dozen uncontrolled fires still raging in the debris of the trade center, with thousands of bodies still buried in the rubble, and with the trauma of the terrorist attacks still fresh in their minds, many New Yorkers were understandably reluctant to return so quickly. Nonetheless, Wall Street and much of lower Manhattan reopened for business on September 17. The nation's top environmental official, Christie Todd Whitman, administrator of the federal Environmental Protection Agency (EPA), who had given her preliminary endorsement of the reopening a few days earlier, issued an official statement of approval on September 18.

"I am glad to reassure the people of New York . . . that their air is safe to breathe and their water is safe to drink," Whitman announced that day.

Similar assurances were given out by the Occupational Safety and Health Administration (OSHA) of the U.S. Department of Labor, which monitors workplace safety, the New York State Department of Environmental Conservation, and the New York City Department of Health.

Even as they made those statements, however, officials knew that their own preliminary environmental tests of the air, dust, and water in lower Manhattan had revealed some troubling readings. The tests found considerable amounts of asbestos and heavy metals had been detected in dust samples throughout the area. Within a few weeks officials would also receive the first results of aerial surveys conducted by the U.S. Geological Survey (USGS) pinpointing the precise locations of hundreds of asbestos "hot spots" on rooftops, buildings, and streets throughout the area, including some that were half a mile or more from the collapsed buildings. Before the end of September, the USGS would also report that dust on the ground and in the air downtown was highly caustic, with alkalinity levels that made it as potent as household drain cleaner. Health officials withheld this information from the public for several months.

Despite their initial safety assurances on September 18, EPA, state, and city officials were still scampering to compile a comprehensive inventory of what contaminants or hazardous materials had been stored inside the mammoth trade center complex before the attacks. They needed the information in order to know what materials were feeding the dozens of fires burning at temperatures as high as 1,000 degrees Fahrenheit and persisting despite all efforts to extinguish them.

Contrary to the public statements, Whitman's agency became so concerned about air safety at its own regional headquarters nearly half a mile north of Ground Zero that between September 13 and September 16 it conducted extensive testing for asbestos in the lobby of the building and on the sidewalks outside. Those tests, using sophisticated equipment that could detect the tiniest of fibers, revealed only one positive result, with asbestos levels of about 25 fibers per square millimeter. This was far below the 70-fiber standard the EPA was claiming to the public was a "safe level" for long-term exposure to asbestos, yet agency officials immediately arranged for vacuum trucks to clean the building's lobby as well as several other offices in the vicinity. They also replaced all heating and air-conditioning filters at their headquarters. The agency told no one of the extensive abatement it had conducted at its own headquarters, and it withheld documents about the cleanup when, in response to a Freedom of Information Act request, it released in late October hundreds of pages on its environmental testing and cleanup efforts.

"There's enough evidence to demonstrate that Mrs. Whitman's statement to the brave rescue workers and the people who live there [in lower Manhattan] was false," said Hugh Kaufman, investigator for EPA ombudsman Robert J. Martin, who launched an investigation early in 2002 into the agency's response to the trade center attack. "She certainly knew it was false by October but never corrected the record," Kaufman said.

Given the scale and unprecedented nature of the World Trade Center catastrophe, it is understandable that during the first few days after September 11 everyone, including public health officials, was focused on guarding against any further attacks and on rescuing the thousands of victims buried beneath the rubble. Surely, no American city has ever confronted a calamity of this scale, nor has any nation faced the simultaneous release of such a complex array of toxic substances into a densely populated downtown area.

Even so, EPA officials and fire-fighting experts were well aware, from previous studies of a handful of spectacular and tragic fires in hotels, commercial buildings, and downtown areas of the United States, that such blazes are capable of releasing a witch's brew of some of the most toxic substances known — including mercury, benzene, lead, chlorinated hydrocarbons, and dioxins.

Despite this prior knowledge, federal officials rushed to dismiss or understate potential health dangers to the public and rescue workers at the site during those first few days. For weeks afterward, even when hope of finding any survivors had long faded, they failed to coordinate or enforce safety efforts to ensure that thousands of firefighters, police, and rescue and cleanup workers at the site were properly protected against toxic releases.

Initially, the various health agencies withheld from the public most results of their environmental testing. The state's Department of Environmental Conservation (DEC) refused outright to release the data, claiming that the test results were part of a "criminal investigation"—presumably the September 11 hijackings—and the city has yet to release all of its data.

On the surface at least, the EPA was more responsive than either the city or state agencies. It began to report some of its test results on its internet web page on September 27. Coincidentally, that was the same day the agency learned that environmental lawyer and activist Joel Kupferman had contacted my newspaper, the New York *Daily News*, and provided us with the results of independent tests he had conducted of World Trade Center debris. Kupferman's results became the first direct challenge to Whitman's all-clear pronouncements. They revealed high levels of asbestos and fiberglass in a substantial portion of the samples. One sample showed 90 percent fiberglass and 5 percent asbestos. From then on, the EPA sought to calm the public by publishing on its web page summaries of daily monitoring reports for asbestos in outdoor air, and the agency eventually expanded those summaries to include the results of periodic tests for more than a dozen toxic substances. As we shall see, the summaries invariably highlighted those results that indicated no danger, while the agency repeatedly downplayed or withheld test results that might raise public alarm.

The federal government has never established ambient safety levels for many of the contaminants that were detected in air samples taken around Ground Zero. Instead of admitting they had no certainty as to what danger these substances might cause, EPA risk experts at the New York regional headquarters devised ad hoc safety "benchmarks" or "removal action guidelines." They then misled the public into believing these were federally approved safety levels and reported that only a few of their test results were above these levels.

Once displaced workers and residents returned to their jobs and homes near the disaster site, a significant number of people began to suffer from respiratory and other health problems. Mark Bodenheimer was one of them. A veteran teacher at Stuyvesant High School, the city's most prestigious public school, Bodenheimer and the rest of the students and staff returned to the building, which is located a few blocks north of Ground Zero, on October 9, when the city's Board of Education reopened the school for classes after conducting a $1 million asbestos cleanup.

"The air in the building smelled terrible," Bodenheimer said. "I had no respiratory problems before this, but I was back there just five days when I started getting constant sore throats and severe headaches." His doctor advised him to get out of the school. Bodenheimer, a Stuyvesant graduate who had taught there for decades, reluctantly accepted a transfer to Bronx High School of Science.

Bodenheimer was no isolated case. A survey of three residential areas near the site, conducted quietly in October by the Centers for Disease Control and the city's own health department, revealed just how widespread such symptoms were: nearly 50 percent of those questioned reported physical problems likely to be related to the trade center collapse, such as nose, throat, and eye irritation, and 40 percent said they were suffering from persistent coughing. Like other disturbing information about the environment around Ground Zero, the public never heard much about this survey. The results were released quietly by the health department in a press release late one Friday afternoon in January 2002— three months after it had been conducted—and received virtually no media attention.

Yet there were too many people getting sick to ignore them all. According to a February 2002 study by the Natural Resources Defense Council, at least 10,000 people in lower Manhattan suffered immediate health problems from exposure to the air near Ground Zero. Faced with a massive public outcry and growing doubts about the environment, federal and local officials hunkered down and kept repeating the same line: any respiratory problems were temporary, a result of smoke and dust from fires that would soon be extinguished. While such symptoms were discomforting, the officials claimed, they posed no serious short-term or long-term dangers.

News of toxic substances other than asbestos being released into the air was not made public until October 26, six weeks after the collapse of the towers, when the *Daily News* published my front-page column on the subject. My information had been gleaned from a quick review of nearly 800 pages of EPA test data, which the agency had been forced to release after the nonprofit New York Environmental Law and Justice Project filed its Freedom of Information Act request. The group's head, lawyer Joel Kupferman, had provided me with copies of all the data.

Only after my article came out did EPA officials concede that their testing had found elevated levels of other contaminants, including benzene, dioxins, PCBs, lead, and chromium, in the air and in water draining into the Hudson River from the trade center. However, agency officials insisted at a City Hall press conference that such high readings had occurred only as occasional "spikes," that they were confined almost exclusively to the immediate vicinity of the debris pile, and that they would soon disappear following the extinguishing of the fires. In short, they posed no danger to the public. As for the water-contamination problem, they contended it was a onetime event after a heavy rainfall and was without long-term consequences.

The fires, however, turned out to be far more difficult to put out than anyone had initially predicted. They burned for nearly four months and even in late January were still smoldering below sections of the debris pile.

The contaminant that got the most attention at first was asbestos, a mineral widely employed as fireproofing material before the federal government banned many of its uses in 1972. Asbestos fibers, once lodged in the lungs, can cause asbestosis, cancer, and mesothelioma, a rare and fatal disease of the lining of the lung. The federal asbestos ban took effect while the Twin Towers were under construction; thus, the mineral was used for fireproofing of steel beams and insulation of pipes in approximately forty floors of one tower and twenty floors of the other. Ever since the ban, the government has regulated removal of asbestos from buildings. EPA rules clearly spell out when and how asbestos must be removed from any public school or commercial building. City and federal officials ignored those regulations at the trade center site. Some critics, including one of the EPA's veteran hazardous-waste scientists and the New York City congressman who represents the area, have since charged that the EPA violated the federal Clean Air Act in its handling of the asbestos problem in lower Manhattan. What seems incontrovertible is that agency officials, in an attempt to minimize people's fears, misled the public about what federal regulations define as a "safety standard" for exposure to asbestos as well as what the legal requirements are for handling asbestos-contaminated matter. In fact, asbestos levels measured in many parts of lower Manhattan were higher than those found in places like Libby, Montana, where the EPA is currently conducting a massive cleanup because of the town's widespread asbestos contamination.

In the case of dioxins, some of the most toxic substances known, the EPA repeatedly told the public that its test results showed very few readings above the agency's "removal action guidelines." In fact, the EPA has no standards for safe dioxin levels in air. Faced with high-level dioxin emissions around Ground Zero more typical of a volcanic eruption or a group of large incinerators packed closely together, the agency's top officials in the New York region simply asked their risk assessors to devise their own thirty-year and one-year removal action guidelines, then told the public that few of its tests had exceeded those guidelines, when in fact a substantial number of them had. EPA scientists in other parts of the country were shocked when they learned that the New York region was posting safety benchmarks for dioxin that had not gone through the agency's normal peer review process.

"Some of us were asking, 'What the heck is Region Two doing?' " one EPA scientist told me.

It wasn't until December that the agency began releasing results of ambient air tests it had conducted for dioxin outside of the actual Ground Zero site. Some of those tests showed high dioxin levels as far as half a mile away from the trade center. Other agency tests showed dangerous levels of PCBs in dust nearly a mile north of Ground Zero, in an area that had been reopened to the public on September 17.

"What happened here is at the level of Watergate," says Dr. Marjorie Clarke, scientist-in-residence at Lehman College in New York and an expert on dioxin and furan emissions from incinerators. "They covered up important information. It just seems to me that, from the get go, a decision had been made from some high-up government types that there is not going to be a problem here."

Federal health and safety officials were not alone in misleading the public, however. Mayor Giuliani, New York City Health Commissioner Neal Cohen, and Joseph Miele, commissioner of the city's Department of Environmental Protection (DEP), abandoned their responsibility to safeguard the public's health and grossly neglected safety issues for thousands of rescue workers at Ground Zero. From the first moments of the attacks, Giuliani assumed direct operational control over all aspects of the governmental response, so much so that, in the daily press briefings that followed for weeks afterward, all governmental officials, from President Bush to New York Governor George Pataki, to the FBI, the EPA, and the Federal Emergency Management Agency, were relegated to secondary and supportive roles behind the mayor. During this period, Giuliani made virtually all major announcements, whether about the death toll, the identities of officials who had perished, the progress of the rescue work, public security procedures, assessments of physical damage to lower Manhattan, traffic restrictions for commuters, assistance to businesses and families of victims, and even dates and locations for funerals of firefighters.

Yet when it came to public health issues and environmental damage, Giuliani and his health commissioner said very little, except to deny any problem whenever the subject was raised by the press, and to reiterate the EPA's assurances about air quality. It seems unlikely that Giuliani and Cohen were simply repeating what the EPA told them out of naïveté. Both men had in the past evinced an arrogant—some would say reckless—disregard for public health matters. From 1999 to 2001, for instance, Giuliani spearheaded a massive pesticide-spraying campaign throughout the city to combat an outbreak of West Nile virus, first with the controversial pesticide malathion, then with the less potent but still dangerous Anvil. The spray campaign, perhaps the largest government urban pesticide experiment in U.S. history, sparked a huge public outcry when hundreds of city residents fell sick from the pesticide fumes and when thousands of fish began to turn up dead in Long Island Sound and in Staten Island's freshwater ponds. Malathion is deadly to marine life and its use is banned over freshwater, a restriction that the city's aerial spraying program had clearly violated.

The following year the city switched to Anvil, but one of the companies selected for massive ground-spraying was fined $1 million by the state DEC for improper training and safety procedures when nearly a dozen of the company's pesticide applicators became ill from their work and filed a complaint with the state. Despite these incidents and accompanying pleas from environmental advocates that the city use other, less toxic methods for mosquito control, Giuliani and Cohen went forward with the spray campaign, and openly ridiculed their critics for unnecessarily alarming the public.

The West Nile virus controversy was only one of several showdowns Mayor Giuliani had with the city's public health community during his two terms in office. Most health advocates, for example, fiercely opposed his repeated attempts to downsize or privatize the New York municipal hospital system, a conflict that led to one of Giuliani's most embarrassing defeats as mayor when the state's Court of Appeals struck down as illegal his effort to sell three city hospitals to a private firm. Health advocates also bitterly and successfully opposed the Giuliani administration's attempt to close twenty-two children's health clinics.

Battles of this sort reflected a philosophical chasm between the mayor and public health advocates. Like many modern-day political leaders who pride themselves on their pragmatic, businesslike approach to governing, Giuliani invariably focused on the immediate and practical results of his policies and paid little attention to long-range repercussions. But public health policy, by its very nature, emphasizes the taking of preventive action today in order to avoid disease and illness in the years or decades to come.

So it should come as no surprise that after the September 11 attacks a legendary hands-on administrator like Giuliani paid so little attention to the public health aspects of the tragedy. Within days of the collapse the various levels of government agreed to a division of labor on safety concerns: City Hall left the responsibility for all testing of the outside air and water around lower Manhattan to federal and state health officials, while it assumed responsibility for checking and certifying the safety of the interior of any commercial or residential areas. The city's portion of the work, in turn, was left to the 6,000-member Department of Environmental Protection, an agency whose primary job is to maintain and monitor the city's vast drinking water and sewage disposal system, but that also has responsibility for handling hazardous-waste problems.

The department, however, did not have nearly enough staff to cope with the pollution hazards it now confronted. Instead of admitting the problem and seeking help from other levels of government, city officials opted for allowing owners of private buildings to carry out their own testing and cleanup with little or no government oversight.

At first, the mass media, especially the local New York City press, dutifully reported the "official story" from the EPA and City Hall and reiterated that all was well with the downtown environment. By the third week after the attacks, however, thousands of people who had returned to work, to school, or to their residences near Ground Zero—even emergency workers at the site—started to complain of serious respiratory problems. It was then that a few journalists, including myself, began to challenge the official story. On September 28, I reported in the *Daily News* that testing of dust samples around lower Manhattan by the New York Environmental Law and Justice Project had revealed more widespread asbestos contamination than Whitman had led the public to believe, and that city officials were not enforcing the wearing of proper safety equipment by the hundreds of firefighters and other workers assigned to Ground Zero.

In a second column, on October 9, I reported that tests conducted by a widely respected environmental firm, Virginia-based H.P. Environmental, had revealed unusually high levels of asbestos fibers inside two office buildings near Ground Zero. A high percentage of the fibers found in those tests were of microscopic size, a result of the enormous pulverization of matter that had occurred from the force of the original collapse of the towers, the firm's scientists told me. The fibers were so small that it was possible they were going undetected by some of the equipment federal agencies were using to detect asbestos.

On October 26, the *News* published a third column of mine, one that was met with a broadside of outrage from the city's political and business elite. That article, accompanied by a blaring front-page headline, "A Toxic Nightmare at Disaster Site," reported that hundreds of pages of the EPA's own documents, which had been released only because of Kupferman's Freedom of Information Act request, revealed a wide array of toxic chemicals being released into the air and water around Ground Zero, sometimes at levels far exceeding federal standards.

Government and business leaders, clearly worried that the report would frighten the public, delay recovery efforts, and endanger property values in the financial district, sought immediately to discredit the column. One of Giuliani's deputy mayors angrily called a top editor at the *News* to complain. The head of the New York City Partnership and Chamber of Commerce fired off a letter, which my paper quickly published, accusing me of "a sick Halloween prank." Even EPA administrator Whitman chimed in with a guest opinion column in the *News* on October 30 to refute my findings.

So great was the backlash that, from then on, top editors at the *Daily News* showed a marked reluctance to pursue stories about environmental pollution downtown, especially when no other newspaper in the city, including the powerful *New York Times*, was following up on our initial reports. One courageous editor at the *News*, however, refused to buckle under the pressure. Metropolitan Editor Richard T. Pienciak not only encouraged my reporting and edited my initial accounts, he simultaneously moved to expand the paper's coverage of the matter. A tall, passionate former investigative reporter who is extremely popular among the rank and file at the paper, Pienciak was openly furious at how quickly our superiors had accepted the assurances of government officials. Having worked on big environmental stories early in his career at the Associated Press, including the infamous Three Mile Island nuclear accident, Pienciak was well aware of how safety agencies often hide the most complex problems from the public. Sensing in his gut that the city faced an environmental disaster, he assigned a special four-person team of reporters to take a closer look at what was happening to residents and office workers in lower Manhattan. Within days of forming the team, Pienciak, who had supervised much of the paper's coverage of the disaster, was removed from his post without explanation. The Ground Zero investigative team he created was immediately dissolved.

I continued to pursue the story in my columns, to the obvious displeasure of the paper's top editors, who took to scrutinizing my copy more carefully than at any time in my fifteen years as a columnist at the *News*. Getting each subsequent piece into the paper became a tense and emotionally draining battle, and they were invariably relegated to the back pages of the paper's news section. Nonetheless, most of them were published, as were several other hard-hitting news stories on the environment by the members of Pienciak's disbanded team, who continued to pursue the issue on their own. As a result, the *Daily News* emerged as the only paper in the city to provide any kind of consistent coverage of the Ground Zero environment.

As for the rest of the New York press, the *Times, Newsday,* the *Wall Street Journal,* and the *New York Post* all accepted at face value the statements from the EPA and City Hall, as well as from a handful of academic experts who lined up to support those agencies, that there was no serious danger to public health. Not until early in 2002, when out-of-town newspapers like the *St. Louis Post-Dispatch,* the *Washington Post,* and the *Los Angeles Times* published major stories on the environmental problems at Ground Zero, did the rest of the New York press begin to pay some attention to the issue.

Meanwhile, a laundry list of chemicals and minerals was being released by the fires and spewing into the streets and buildings around Ground Zero, the underground network of subways and utility conduits beneath Manhattan, and the nearby Hudson River.

To understand the enormity of the environmental problem we need to come to grips with the sheer size of what was destroyed on September 11 — the equivalent of a small American city. The quantity of contaminants contained within the buildings is truly staggering.

Take just one substance, lead, as an example. Lead is an extremely dangerous heavy metal. Inhaling even minute quantities of lead dust over an extended period can cause brain damage. It is especially dangerous to children who sometimes breathe lead dust or eat chips of lead-based paint flaking from household walls. The use of lead in paint has been banned in the United States for decades but the interiors of many inner-city tenements still contain undercoats of it. At the trade center, the danger came not from lead in paint but from lead inside computers. The average personal computer contains anywhere from four to eight pounds of lead. We know that approximately 50,000 people worked in the two World Trade Center towers, and that most of them used personal computers. Several thousand more worked at Seven World Trade Center, a forty-seven-story building just north of the Twin Towers, and at other, smaller structures on the site that were also destroyed. We can thus assume that at least ten thousand PCs, in addition to hundreds of servers and mainframe computers connected to them, were pulverized into dust that day or vaporized by the fires in the subsequent months. It is likely, therefore, that a minimum of 200,000 to 400,000 pounds of lead were released into the air, the ground, and the buildings around the site.

Despite all the assurances of the governmental agencies, and despite the uncritical backing the mass media gave to the initial safety pronouncements, it was impossible to conceal the fact that thousands of people who had returned to live or work in downtown Manhattan were getting sick from breathing noxious air. An unusual number were having their first ever asthma attacks. Others found themselves with frequent nosebleeds. This was not happening in some poor, out-of-the-way neighborhood where the complaints of residents could easily be dismissed. It was occurring in neighborhoods like Tribeca and Battery Park City, affecting upper- and middle-class residents, students at Stuyvesant High School, stockbrokers and lawyers, and even government workers in the financial district.

The toxic releases, as I will show in the following pages, amounted to more than mere "spikes." In some cases, persistently elevated levels of some toxins were found by government testers, and not just in the rubble pile at Ground Zero. Only months later, in early February 2002, did independent air-monitoring tests from the University of California at Davis report that air pollution levels in lower Manhattan during the previous October had been worse than during the oil fires in Kuwait after the Gulf War. The UC Davis team recorded those levels at a monitoring station one mile north of the trade center site, a station that was not even in the path of prevailing winds.

Even if all individual contaminants in the air had been below permissible federal safety levels, there is yet another troubling concern for many scientists, what some call the unknown synergistic effect of exposure to even low levels of a variety of toxic substances at one time. "There were probably a thousand or more chemicals in that soup," says industrial hygienist Monona Rossol. "No one knows how that could affect a person."

In the following pages I attempt to recount the details of the massive toxic deception that took place, to document the frustrations of the people who were affected, and to explore some vital questions that have yet to be addressed. What were the chemicals released by the trade center collapse and where did they go? What did EPA representatives and other government officials know about the possible extent of contamination and when did they know it? Why did the EPA tell the public one thing about the health risks from asbestos exposure in New York City, yet tell a quite different story, and apply a far tougher safety standard, for asbestos removal in other parts of the country? Why did the federal government and the Giuliani administration neglect for many months two of the most important health issues of the entire catastrophe: proper safety protection both for rescue and recovery workers and cleaning up the fallout inside hundreds of high-rise office and residential buildings around the site?

The mass media must share the blame for this deception. Most press organizations in the city displayed a woeful ignorance when it came to interpreting and explaining the environmental issues at stake. In the first few days after the attack it was understandable that journalists would be reluctant to raise troubling questions. The country, after all, was facing an unprecedented crisis. But as days turned into weeks, and as the disaster site turned into a huge debris-removal operation, there was no excuse for the failure of safety officials to make sure that all firefighters and emergency workers were

properly protected from dangerous chemicals—or for journalists not to accurately report those failures.

Today, more than 200 New York City firefighters who
served at Ground Zero are on medical leave, and as many as
700 have exhibited respiratory problems, what is now called
the World Trade Center Cough. Many of those have been
assigned to light duty, and it is feared a good portion may
never be able to fight fires again.

In addition, a troubling number of rescue workers from
other parts of the country who had volunteered at Ground
Zero and then returned home are reporting serious health
problems.

In Ohio, thirty-seven of seventy-four members of Ohio
Task Force One, a group of emergency responders who volunteered to work at Ground Zero, have become ill since re-
turning home.

In California, 100 of 395 emergency responders who
worked at Ground Zero between September 12 and October
7 have filed workers' compensation claims because of illness
they say is related to the World Trade Center catastrophe.

Five months after the disaster, Dr. Stephen Levin of the
nationally known Selikoff Center for Occupational and Environmental Medicine at Mount Sinai Hospital stated that a
"high percentage" of the hundreds of iron workers and other

recovery personnel at Ground Zero that his center has examined have experienced respiratory problems.

Experts who have carried out long-term studies of the health effects of such fires suggest that this is only the tip of the iceberg of the health problems firefighters and rescue and cleanup workers will face in the future.

Christie Whitman, Mayor Giuliani, and the other public officials should have told New Yorkers the truth from the start—that no one could guarantee the air around Ground Zero was safe because no one had ever confronted a disaster of such proportions. They should also have released all the raw data on government testing as soon as they had the results and made clear that safety levels for many of these toxins did not even exist.

The early blanket assurances that government officials issued were a grave mistake, and their continued defense of those assurances in the face of widespread public skepticism was inexcusable. Thousands of people may end up paying for that deception through unnecessary illness or premature death in the decades to come. In their rush to return New York City and Wall Street to business as usual these short-sighted officials unwittingly paved the way for a second wave of victims from the World Trade Center tragedy.

In scrutinizing the government response to the collapse I

am not suggesting there was some secret conspiracy to hide the facts, or that anyone intentionally set out to mislead the people of New York. But I have no doubt that some officials saw what they wanted to see and irresponsibly chose to minimize or dismiss any post-collapse environmental threat. They did so for any of a number of reasons or combination of them: out of ignorance; in response to political pressures from superiors; or because they underestimated environmental threats from toxic releases, something that is all too common in the history of such disasters. Once ordinary citizens questioned their assurances, however, those officials closed ranks, dissembled, hid important information, and refused to listen or alter their policies. My aim here is to outline key lessons of this saga for the general public so that this kind of deception is not repeated should another catastrophe occur in the future.

As in any tragedy, there are always heroes, courageous individuals who refuse to be silenced, who persist in challenging the official story, who doggedly press forward to uncover the facts, no matter how much they are chastised or ridiculed by those in power. The heroes of this story are people like lawyer Joel Kupferman and industrial hygienist Monona Rossol; scientists like Cate Jenkins of the EPA, Piotr Chmielinski of H.P. Environmental, and Tom Cahill of UC Davis; community leaders like Marilena Christodoulou, president of the Stuyvesant High School Parents Association, and Maureen Silverman, of the World Trade Center Emergency Environmental Group; journalists like Richard Pienciak of the *Daily News* and Andrew Schneider of the *St. Louis Post-Dispatch*; and political leaders like U.S. Representative Jerrold Nadler and U.S. Senator Hillary Clinton. Each, in his or her own way, listened to the voices of people living or working in lower Manhattan; each refused to be lulled by the failures of most of the New York media to report what was actually happening; each helped to pierce the false claims of government officials and to validate the concerns of rescue workers and ordinary citizens, to confirm to them that they were not imagining their ailments. Perhaps the combined efforts of these voices will soon force the various government agencies to assume their proper responsibility for a complete environmental cleanup of lower Manhattan, something they should have done from the start.

1

Day Turned into Night

SEPTEMBER 11, 2001, began as a warm, cloudless day, with brilliant sunshine blanketing the nation's east coast. The splendid weather was especially welcomed that morning in New York City, where polling places had opened at 6 A.M. so voters could cast their ballots in a hotly contested municipal primary to select the Democratic and Republican candidates who would vie in November to succeed Rudy Giuliani as mayor. A no-nonsense former federal prosecutor and New York's first Republican mayor in nearly a quarter century, Giuliani was nearing the end of his second term and was prevented by term limits from running again. In a post often called the second toughest in America, he had won national accolades for sharply reducing both the city's crime rate and its welfare rolls and for presiding over its downtown economic renaissance. But after eight years of his regime, even some of his admirers had become weary of his combative and tyrannical style, and of the way his personal life was turning into an embarrassing public soap opera of marital infidelities and courtroom scandal. His detractors—with Giuliani, either you love him or hate him—eagerly awaited his departure. He had especially incurred the enmity of many in the African-American and Hispanic communities, which together make up more than half the city's population. Those groups felt left out of the economic boom of the 1990s, and they believed that the police department, with Giuliani's direction and blessing, had engaged in rampant racial profiling and unchecked violations of civil rights. For them, as well as for many of the former Giuliani faithful, the September 11 mayoral primary represented a chance for a new beginning.

By 9:30 A.M. a day that began with so much brightness and hope had turned into one of the deadliest in the nation's history. By then the world had been stunned by the news— much of it captured live on television—that terrorists had hijacked four passenger airplanes, then turned them into deadly missiles, methodically crashing one into each of the twin towers of the World Trade Center, and another into a wing of the Pentagon in Washington, D.C., before the fourth finally plummeted into a field in western Pennsylvania. As city residents tried to digest what had happened, the two 110-story trade center towers—hallmarks of New York City's skyline and international symbols of American power for decades—became engulfed in raging fire and smoke, with thousands of people still trapped inside, many of them desperately jumping to their deaths from windows to escape the flames. Barely an hour after the second attack came the final horror: the astonishing collapse, first of one, then another of the two towers. As each building crumbled it released a torrential dust cloud of glass, concrete, steel, and other debris into the atmosphere. So thick and impenetrable did those clouds become that they blotted out the sun, turning day into night. Thousands of office workers who had fled from nearby skyscrapers into the narrow streets of the financial district, and who had then stopped to stare incredulously at the burning buildings, found themselves trapped inside the dust cloud, which whipped toward them with the force of a hurricane, as flying glass and debris rained down from the sky.

Lawyer Marcie Murray evacuated her office at 120 Broadway after the second plane crashed into the South Tower at 9:03 A.M. She rushed out into the street along with her paralegal assistant and a new client with whom she'd been meeting. Joining countless others, Murray milled about at Nassau and Cedar Streets, about two blocks southeast of the trade center, as she waited for firefighters to contain the blaze so she could go back to work. From her spot she had no direct view of the flames, so when she heard a deafening boom just after 10 A.M., she had no inkling that it was one of the towers collapsing—until she saw hundreds of people running toward her in panic with a tornado of dust chasing them.

"We tried to outrun what turned out to be Godzilla, but couldn't," Murray recalled. "We had been running for our lives and were short of breath when we stopped and were stopped—by blindness, panic, and that dark gray stuff. We gulped down huge quantities of that raw avalanche. For the next couple of days every millimeter of my skin felt like I had lain down with the cacti. My husband, a contractor, immediately identified it as fiberglass. And that was on my outsides."

Police and firefighters somehow managed to evacuate most civilians from the devastated area, but by then several square miles of lower Manhattan were enveloped in the dark, thick, acrid cloud. Those who escaped, many of them dazed and caked from head to toe in fine white dust and grime, formed a human stream pouring across the East River over the nearby Brooklyn Bridge; others scampered northward for several miles until they reached the safety of midtown.

The stunned firefighters and police who managed to survive the initial collapses—more than 300 of the first 400 fire-

fighters to arrive on the scene perished—were soon re-
inforced by fresh waves of emergency personnel from all
over the metropolitan area. The newcomers launched a
frantic search for victims buried in the rubble. They wrestled
to contain the raging fires as well, but with little success. By
late afternoon, flames had spread throughout the sixteen-
acre site, especially below street level where the trade
center's huge seven-story concourse area of restaurants and
shops was located. Eventually, the blaze reached all the way
to Seven World Trade Center, just north of the main complex, where fire was reported at 4:10 P.M. It was inside that
building, ironically, that Mayor Giuliani had constructed a
multimillion-dollar state-of-the-art emergency command,
which was supposed to be the city's nerve center in case of
just such a catastrophe. By 5:20 P.M., just over an hour after
the flames were first reported, the entire forty-seven-story
building collapsed.

It would be months before the public learned anything
about what caused the fire at Seven World Trade Center or
the building's astonishingly rapid disintegration. Several
thousand gallons of diesel fuel, which had been stored in
emergency tanks below street level to provide power to the
city's command center, ended up feeding the flames and
melting steel girders on which the building rested. The collapse, in turn, destroyed two Con Edison electrical substa-

tions located at street level, over which the building had been constructed. The substations, the nexus of a huge power grid, were the terminal point for several high-power voltage lines, including one that ran several miles from Con Ed's Farragut power plant in Brooklyn, under the East River, and across lower Manhattan. Inside, nine forty-foot transformers and scores of smaller capacitors divided electricity into smaller feeder lines that supplied power to the trade center and other downtown buildings.

Not until several months after the attack did Con Edison publicly reveal—and then only after my repeated inquiries—that the collapse of Seven World Trade Center had led to the spilling of 130,000 gallons of oil, most of it contaminated with low levels of PCBs, when the transformers and the Farragut line ruptured. The spilled oil, in turn, helped fuel underground fires that burned below Seven World Trade for several weeks. Those fires contributed significantly to the release of high levels of cancer-causing dioxins and furans into the air; an EPA air-monitoring station set up less than a block away at West Broadway and Barclay Street detected high levels of both substances from September 16 into early November.

In the weeks after the collapse the public learned nothing about the spilled oil or the ruptured diesel tanks from the EPA, the city, or the state. Con Edison did report the spill to the state Department of Environmental Conservation within hours of its occurrence, as required by federal law, so government officials knew they had a major problem on their hands almost immediately, even if they did not know the size of the release or what had become of the oil. Yet the various agencies chose to say nothing publicly about the spill even as they gave the go-ahead to businesses and residents to return to the buildings around Ground Zero.

The magnitude of the destruction that occurred that day is still difficult to grasp. It was not just a matter of two of the world's tallest buildings crumbling within a few hours. In addition to Seven World Trade Center, two other buildings on the site, Five World Trade Center and the Marriott Hotel, were completely destroyed. Three others, One Liberty Plaza, Four World Trade Center, and Six World Trade Center, were partially destroyed, and nine buildings adjacent to the site suffered major damage. Several subway and Port Authority commuter tunnels and stations, a six-level parking garage, and the entire trade center concourse were also destroyed. A vast rubble pile estimated at 1.2 million tons of steel and building material was all that remained.

The area around the trade center site is one of the most densely populated on earth. Several hundred thousand people commute into lower Manhattan each day to work in scores of high-rise buildings within a mile radius of the site.

A few blocks from Ground Zero are situated City Hall, the federal courts, most major federal, city, and state agencies, and the New York Stock Exchange and other major financial markets, as well as the headquarters of some of the nation's biggest banks. In addition, 34,000 people lived in the sprawling Battery Park City complex, the Beekman Hospital area, Tribeca, and Chinatown, all neighborhoods within a mile and a half of the site. Many residents of those neighborhoods also worked in the financial district. They enjoyed the advantage of being able to walk to their jobs each day as well as the vibrant restaurant and nightlife that had mushroomed in the area over the past decade. They were generally affluent, well-educated communities, people with a high level of confidence in official government pronouncements.

During the first week after the attack, government officials closed off all of lower Manhattan below Canal Street as they scrambled to save victims, contain the fires, and assess the damage. The vast majority of local residents were displaced from their apartments by the flames and smoke or by the loss of power. At places like Battery Park City, the closest neighborhood to the site, the entire complex was evacuated.

By September 17, however, government leaders gave the go-ahead for the financial markets and all businesses east of Broadway or north of Canal Street to reopen. Automobile traffic remained barred from the entire area, and many bridges and tunnels into Manhattan stayed closed, but over the next several weeks the zone around Ground Zero that was designated off-limits to the public kept shrinking.

EPA administrator Whitman issued two press releases, on September 13 and on September 18, to reassure New Yorkers that there was no apparent environmental danger from the collapse or the fires.

"Monitoring and sampling conducted on Tuesday [September 11] and Wednesday [September 12] have been very reassuring about potential exposure of rescue crews and the public to environmental contamination," Whitman said on September 13.

Her statement was backed up the following day by John L. Henshaw, assistant secretary of labor and head of OSHA.

"Our tests show that it is safe for New Yorkers to go back to work in New York's Financial District," Henshaw said in a press release.

Henshaw, whose agency is in charge of protecting worker safety, went even farther out on a limb. His words sounded like a federal certification that the insides of downtown buildings were safe as well.

"New OSHA data also indicates that indoor air quality in downtown buildings will meet standards," said the first paragraph of Henshaw's release. Buried way down in the text of the release, the agency conceded that its own tests of dust samples outside of buildings and on cars and other surfaces in lower Manhattan had detected 2.1 to 3.3 percent asbestos levels—above the 1 percent level that the federal government defines as hazardous asbestos material. But the agency stressed that air tests OSHA had conducted showed that ambient air quality "is not a case for public concern."

The initial assertions from Whitman and Henshaw, as we will soon see, were at best incomplete and at worst misleading, for they were based on only a small number of tests, most of them for asbestos. By September 18, results of more extensive tests had come in. They revealed a bigger asbestos problem and a more troubling range of toxic emissions, many of which the EPA did not even have safe standards for.

The enormity of the disaster, and the way dust and debris from the collapse of the buildings spread over such a large area (some was discovered as far away as Brooklyn), should have told officials from the start that it was not possible within a couple of days to predict with any certainty whether the environment was safe. During the first week, city officials did not even have an inventory of what toxic substances had been stored inside the trade center before its collapse.

A few days after the disaster, for example, Walter Hang, who runs Toxics Targeting Inc., an Ithaca-based consulting firm that tracks information on hazardous materials for governments and utilities around the nation, received a call from an official at New York City's Department of Design and Construction.

"The city and Port Authority, the state agency that had built and managed the complex for decades, didn't know what hazardous materials might be feeding the fires," Hang said, "so they asked me to produce a fast report on all known toxics stored by any World Trade Center occupants." Hang conducted an immediate search of more than a dozen different local, state, and federal computer databases where hazardous materials and spills are routinely reported. By the time he'd finished he had nearly a hundred pages of information. He turned his report over to the city on September 18.

It revealed that there were more than half a dozen diesel fuel storage tanks, with a total capacity of more than 70,000 gallons, located in the lower levels of various buildings in the complex. The tanks, owned by the Port Authority, the city, Salomon Smith Barney, and Verizon, had been placed there to supply emergency power to computers or telephone switching systems in case of electrical blackouts. In addition, Hang discovered that the U.S. Customs Service had a large laboratory on the eighth floor of Six World Trade Center. In the past, the lab had reported storing thousands of pounds of arsenic, lead, chromium, mercury, and other toxic substances for its work. But the amount of hazardous materials stored by some of the trade center's occupants does not begin to compare in magnitude to the enormous array of toxic substances that were contained inside the construction materials of the buildings, and especially as part of the office equipment and furniture those buildings housed. We will look at this catalog of poison in more detail in chapter three.

2

Ignoring the Lessons of History

TO DATE, no official from any government safety agency has publicly acknowledged the dimensions of the toxic releases that resulted from the World Trade Center disaster. To fully understand how these agencies failed properly to protect the public, we must first summarize what health officials knew—or should have known—about the environmental threat posed by these unprecedented events.

The EPA's Region Two headquarters is located at 290 Broadway, in a federal building about half a mile northeast of the disaster site. The agency's own records reveal that on the day of the attack no EPA emergency team took samples of the air in lower Manhattan. Two teams did conduct a handful of air tests that day for asbestos, lead, and volatile organic compounds, but they did so outside of Manhattan. One team took three air samples at Liberty Park in Jersey

City, New Jersey, and another collected four samples in downtown Brooklyn, where the wind was driving the huge plume of smoke.

While it is understandable that during the chaos of the first few hours after the attacks agency staff might have found it difficult to put together a rapid response field team, the fact that neither the EPA nor any local health agency obtained samples close to the source of the colossal dust clouds of September 11 means that no one will ever know the precise level of toxic contamination, or the combination of toxins, that enveloped tens of thousands of people that day. We do know the cloud was so thick that people caught in it could not see their own hands and could barely breathe for several minutes afterward. Many reported swallowing thick wads of dust in their desperate efforts to take in air.

While the scale of the collapse was unprecedented, this was not the first time health officials had dealt with a major commercial building fire or uncontrolled burning of toxic chemicals, so there certainly existed prior experience by which to judge the potential dangers to firefighters, emergency workers, and the general public. Among famous past fires whose effects experts have studied are the 1977 New York Telephone Company fire, the 1980 Las Vegas MGM Grande Hotel fire, and a 1978 blaze at a toxic waste dump in Chester, Pennsylvania. The lessons learned from these and other similar catastrophes should have served as a warning to those in charge of Ground Zero that the collapse of two of the world's tallest buildings in the midst of a densely populated city, followed by months of uncontrolled fires, was an environmental disaster with probable long-term health impacts. A frank disclosure to the public of potential dangers and the immediate implementation of steps to minimize any such long-term damage to the local population were urgent necessities that officials failed to carry out.

LESSONS FROM PREVIOUS FIRES

Professional firefighters and construction engineers have known for years that the fumes and smoke from fires in modern office buildings are far more dangerous and intense than those involving products like wood, paper, and natural fibers. This is due largely to the widespread use of heavy metals and synthetic chemicals in the manufacture of products such as personal computers, urethane foam and styrene furniture, nylon carpeting, polyvinyl chloride–insulated computer and telephone cables, as well as bromine in fire-retardant products.

"Rigid polyvinyl chloride [PVC] emits 60 percent of its weight as hydrogen chloride, HCL, in the early stages of a fire, which coats soot particles," notes Dr. Roderick Wallace in a recent study that compares the trade center disaster to previous fires. "These particles are small—smaller than those given off by, say, wood or cotton. They get into the lungs and deliver a huge dose in a short time, destroying lung tissue by sheer corrosion. . . .

"Because of the flame retardant properties of chlorine," Wallace writes, "much more mass of the PVC goes off as soot and the smoke is extremely dense. People cannot see to escape. Plasticized PVC, as used in communication cable, also produces highly acidic soot and fumes of dense smoke, but also gives off masses of phthalate and anhydride, very irritating and explosive. Both forms of PVC produce large quantities of benzene, xylene, and toluene which are narcotic, and may give rise to secondary explosions. . . . PVC fumes and soot may contain as many as 300 species . . . ranging from formaldehyde to the polyaromatics, with and without chlorine."

In addition, some nitrogen-based plastics release highly poisonous hydrogen cyanide (HCN) shortly before ignition. The toxic chemicals released by the burning of these synthetic products can cause far more permanent injuries and deaths in fires than the actual flames.

In 1980, for instance, fire broke out in the ground level of the MGM Grande Hotel in Las Vegas, Nevada. Rising smoke from burning plastics trapped many of the hotel guests in their rooms, killing eighty-five of them at a considerable distance from the actual flames. After the fire was extinguished, fifty-four of the injured survivors participated in a follow-up study as part of legal suits against the hotel. Among the chronic problems those survivors experienced, according to Wallace, were:

Frequent sore throats (51 percent)
Shortness of breath (65 percent)
Phlegm production (50 percent)
Hands/feet fall asleep (68 percent)
Headaches (59 percent)
Memory lapses (56 percent)
Change in perception abilities (54 percent)
Dizziness (47 percent)
Acnelike outbreaks or rashes (22 percent)
Depression (70 percent)

In addition, among women in the group who were of reproductive age, 84 percent experienced new menstrual difficulties.

The survivors of this hotel fire were in all probability exposed to fewer toxic contaminants than were the tens of thousands of people made vulnerable by the trade center collapse on September 11. Yet, to date, the federal government has done nothing to create a comprehensive medical database of those people who were enveloped by the dust clouds that day. Only a small study initiated by Columbia University of pregnant women who were caught in the cloud has received government funding. In the case of civilians, who are not trained or equipped to handle a catastrophe such as a fire or massive collapse, the horror of those events can also create significant trauma and lasting psychological effects. The lesson of the MGM hotel should serve as a warning that, given what they went through, many of the people caught in the initial trade center collapse will need long-term medical and psychological monitoring, yet this vast group has been virtually ignored by governmental institutions.

Likewise, the even greater dangers faced by the firefighters and rescue workers at Ground Zero have been poorly and haphazardly addressed. Again, there are plenty of historical lessons to draw from.

In 1977 a fire erupted at a New York Telephone Company building in lower Manhattan that contained about 100 tons of plastic PVC cable. It took 700 firefighters more than fourteen hours to extinguish the blaze. More than 200 were injured and about half of those subsequently participated in a 1980 study sponsored by the Uniformed Firefighters Association. Among the senior firefighters, those who had been on the force for fifteen years or more, an astounding 80 percent reported permanent injury.

Among the long-term effects were: impaired disease resistance (37 percent), coughing (33 percent), hoarseness (23 percent), loss of lung function or pain (15 percent), asthma or upper respiratory allergies (15 percent), growths, epidermal or membrane (14 percent), sensitivity to smoke (11 percent).

The 200,000 pounds of PVC that burned in the telephone company are "a minor event," says Dr. Wallace, when compared to the toxins produced at the trade center.

Perhaps the most chilling example of the profound danger posed by fires that involve massive amounts of synthetic chemicals is the 1978 fire at the old Eastern Rubber Reclaiming Inc. plant in Chester, Pennsylvania, fifteen miles south of Philadelphia. The plant, known to local residents as the Wade dump because it had served for years as an illegal warehouse for discarded barrels of chemicals, exploded into flames on the morning of February 2. More than 200 firefighters, police, and emergency personnel fought the blaze for a day and a half. Nearly twenty years later, a team of investigative reporters from the *Philadelphia Inquirer* began tracking down those firefighters to find out what had happened to them. The results of that investigation, published in a series of articles in the spring of 2000 by the *Inquirer*, are a tragic blueprint of what awaits many of the firefighters and rescue workers who responded to the World Trade Center fires.

According to the *Inquirer*'s exhaustive study, within the first ten years after the blaze, "at least 21 cancers alone were diagnosed among men who worked the fire or its aftermath—men who were, on average, in their mid-40s. Eleven had died. Since then, 18 new cancers have appeared and killed nine more."

Ten of those men died of lung cancer, nearly five times the rate of the general population. Melanoma, the deadliest form of skin cancer, was found at six times the expected rate. Two men who had walked side by side all over the site collecting chemical samples while the fire was burning ended up dying within fourteen months of each other from amyotrophic lateral sclerosis (ALS), popularly known as Lou Gehrig's disease. Normally ALS strikes two out of 25,000 people.

The *Inquirer* found that "a vast assortment of other malignancies, blood ailments and vascular disorders also occurred. The list includes four colon cancers, three prostate cancers, two brain cancers, two liver cancers, two kidney cancers, three cases of peripheral vascular disease, and single cases of Hodgkin's disease, aplastic anemia, leukemia, and cancers of the pancreas, esophagus, larynx, thyroid and chest wall." Several relatives of the men who died reported that the men had started to experience medical symptoms shortly after fighting the blaze.

One such was firefighter Stump Swanson. After rupturing blood vessels from coughing so hard he checked into a medical center for tests in November 1979. Within ten days, at the age of forty-two, he was dead of lung cancer. Another victim was Chester Fire Captain Marvin Cherry. He and the company he commanded were sent into the dump after the main fires had been extinguished to collect the fire-fighting equipment that had been left at the site. He returned twice afterward to fight smaller fires that had erupted there. Three years later, in early 1981, he started to cough and wheeze incessantly. By then he had lost fifteen pounds, and when he visited his doctor to find out what was wrong he discovered he had lung cancer. The disease quickly spread to his brain, skeleton, and liver, and Cherry, a robust and healthy man until that time, was dead by December, at the age of forty. During the final months of his life, Cherry became convinced that the fire at the Wade dump was the cause of his disease.

Many of the same chemicals that burned at the New York Telephone Company fire and at the Wade dump were re-

leased in the World Trade Center fire in far greater quantities—toluene, mercury, phenols, lead, xylene, chromium,
zinc, vinyl chloride.

Unfortunately, in the early days after the collapse, as we
shall see, officials in charge at Ground Zero failed to fully
grasp the immense health risks emergency workers faced
and did not provide those workers with sufficient protection.
Nor did federal health officials honestly convey to the public
how little they knew about the safety situation for those re-
turning to their homes and jobs in the area.

3

Anatomy of a Toxic Nightmare

JUST WHAT WERE the toxic substances released by the collapse and the fires? What do we know about the size of those releases or their potential threat to human health? What did the EPA and OSHA discover in their own environmental testing and what did they report to the public or withhold from it? There has been much confusion and a good deal of misinformation dispensed about these issues since September 11. Some of the confusion is understandable, given the enormous number of chemicals released and the relative ignorance among ordinary citizens about the effects of such releases on human health. In addition, the federal government has standards for only a finite number of toxic substances, and often these have been devised more for occupational exposure than for exposure in ambient air.

What follows is my own assessment of the extent of contamination, potential and real, that occurred on September 11 and in the weeks that followed. It is based on scores of discussions during the six months following the event with government health officials, scientists, industrial hygienists, experts in commercial office equipment, and environmental safety advocates, on reviews of thousands of pages of monitoring reports of the various agencies involved, on testimony in various public hearings that have addressed the catastrophe, and on interviews with local residents and victims of the attack. The list is only a preliminary sketch; it does not deal with every major contaminant that was released, such as fiberglass or sulfur dioxide or freon, but focuses instead on those substances that may end up causing the greatest long-term health effects, and especially on those about which the public has received incomplete or misleading information from government officials.

ASBESTOS

As mentioned earlier, asbestos received the most attention from health officials and the press, even though it was only one of hundreds of hazardous substances present in building materials at the Twin Towers. Estimates of how much the two skyscrapers contained vary from 400 to 1,000 tons. Such estimates do not include the asbestos insulation around the labyrinth of steam pipes and other utility conduits that were destroyed beyond the immediate vicinity of the towers, or the asbestos contained in other buildings in the complex that suffered major damage.

To put that amount in perspective, we would do well to compare it with another major asbestos disaster in New York City more than a decade earlier. Around 6:30 on the evening of August 19, 1989, a twenty-four-inch underground steam pipe exploded in the Gramercy Park section of Manhattan, leaving a ten-foot-wide crater in the middle of East 20th Street. Three people were killed, twenty-four were injured, and thousands of residents were evacuated as a roaring geyser of scalding steam shot high into the air for hours, blanketing the street and nearby buildings with a thick layer of mud and debris. Con Edison, which owned the pipe, did not reveal until four days later that the mud was contaminated with asbestos, by which time residents had already returned to their apartments and begun their own cleanup. Tests by city officials found no elevated levels of asbestos in the air, but fearing that any disturbance of the asbestos-laden dry mud and dust would make the fibers airborne, the city again evacuated 350 people from five buildings, cordoned off the area, and ordered an extensive cleanup of each apartment. The work took more than seven months, with Con Edison spending $90 million for both abatement and compensation to residents who lost property. The company and two top officials were eventually indicted and convicted on criminal charges of lying about the asbestos contamination.

The astonishing aspect of this story is the amount of asbestos that created the furor: only 200 pounds were released in the Gramercy Park tragedy. The simple presence of asbestos in mud and dust (remember, none was detected in the air) triggered a massive seven-month cleanup.

Compare the reaction of health officials in that incident with the World Trade Center collapse, which involved up to 1,000 tons, and where high levels of asbestos were detected in the open in dozens of the first monitoring tests, as well as in a high percentage of dust samples from all over lower Manhattan, and you begin to understand how the EPA and other agencies simply turned their backs on the problem. To avoid a massive government-financed asbestos cleanup of lower Manhattan, say critics inside and outside of the EPA, the agency's top brass effectively violated federal law and their own regulations on asbestos removal. To justify their inaction, as we will see in the next chapter, they lied to the public about what constitutes a dangerous level of asbestos, and they then refused for months to address the problem of asbestos contamination inside many downtown Manhattan buildings.

LEAD

As mentioned previously, anywhere from 200,000 to 400,000 pounds of lead were present in the thousands of personal computers that were instantly destroyed when the Twin Towers crumbled to the ground. An unknown but significant quantity of lead was also inside thousands of batteries, on countless electrical soldering connections, in water and steam pipes throughout the complex. Much of that lead became pulverized into microscopic dust by the enormous force of the collapse. When released into the air, lead easily attaches itself to other particulates and can travel long distances before settling to the ground. Pure lead does not break down, but lead compounds can be changed by sunlight, air, water, or fire. Once on the ground, the metal can migrate into bodies of water.

Exposure to lead can damage almost any part of the body, especially the central nervous and reproductive systems and kidneys. Most at risk are infants and unborn children, who can suffer developmental problems and brain damage. In addition, some lead compounds have been declared carcinogenic by the U.S. Department of Health and Human Services.

According to EPA records, between September 16 and October 2, the agency collected thirty-four air samples for lead in the area immediately around the trade center. Six of those samples, or 17.5 percent, showed lead levels above the federal safety standard of 1.5 micrograms of lead per cubic meter of air. When I reported the positive findings, health officials referred to these and other high readings of contaminants as temporary "spikes" or as "snapshots in time" that posed no long-term health concerns. In the case of lead, however, the federal standard is based on only three months of continuous exposure to those levels. More important, these and many other tests the agency conducted were "grab samples," i.e., a specific quantity of air was collected for a short duration of time and then analyzed, as opposed to testing for an extended twenty-four- or seventy-two-hour period. So it is entirely possible that in other spots that were not tested, or even in the same spot at a different hour of the day, the lead levels were higher. The fact that a significant percentage of the agency's tests were showing lead levels above federal safety standards was at least cause for concern. True, those elevated levels were found only within a "red zone," which at the time was off-limits to all but rescue and recovery workers, but that was because the agency was conducting virtually all its tests for heavy metals within that red zone. Only months later would New Yorkers learn that a twenty-four-hour air-monitoring station set up by a team of scientists from the University of California at Davis had registered elevated lead levels in the air a mile north of Ground Zero.

Furthermore, even if all the lead results had been low or nondetectable, no one had any inkling of how much lead was in the original September 11 dust clouds. Pregnant women exposed to those massive dust clouds, for instance, should have been given an immediate advisory about possible lead exposure. When lead first enters the body it migrates into the blood, but within two weeks it gets deposited in the bones.

"In the case of pregnant women, you want them to up their calcium since the fetus in the woman's body can't tell the difference between calcium and lead," one scientist told me. "If you give the mother a big dose of calcium immediately the fetus will up its intake of the calcium instead of the lead. But I guess they [government officials] were afraid of upsetting everyone with advice like that."

MERCURY

Mercury is another heavy metal that is extremely toxic to humans and especially dangerous to children, since it accumulates in the body.

Short-term exposure to high levels of metallic mercury

vapors can cause lung damage, nausea, vomiting, diarrhea, increases in blood pressure or heart rate, skin rashes, and eye irritation, according to the federal Agency for Toxic Substance Disease Registry. Organic or inorganic mercury can also damage the brain, the kidneys, and a developing fetus. In the brain, effects can result in irritability, tremors, changes in vision or hearing, and memory problems.

For decades mercury has been routinely used in thermometers and thermostats, power and telephone switching systems, batteries, liquid crystal computer monitors, and one of the most common fixtures of the modern building, fluorescent bulbs. The average four-foot fluorescent bulb contains about 21 milligrams of mercury, though a few environmentally engineered models use considerably less. That is a tiny amount that would fit on a pencil dot. Nonetheless, "there is enough mercury in any lamp to become a serious environmental problem if it is released uncontrolled," according to Paul Abernathy, a California businessman who recycles mercury.

The mercury contained in twenty-five lamps can pollute a twenty-acre lake. It is so toxic that all discarded lamps are considered hazardous waste and cannot be disposed of in landfills or incinerated, but must be properly recycled or sent to specific dumps.

"Any lamp has more mercury in it than you want to

breathe," said Abernathy, who has testified as an expert witness in numerous court cases involving mercury releases. "If you break thousands of them all at one time, you have a significant release of mercury."

There were 500,000 fluorescent lamps inside the Twin Towers on September 11, according to the Port Authority of New York, and unknown thousands more in the other buildings that were damaged or destroyed. Every one of those bulbs shattered when the buildings came down. Whenever a fluorescent bulb shatters the mercury breaks up into globules, most of it vaporizing within eight days, or sooner if there is heat present. The vapors, once they cool, easily attach themselves to other particulates. A Port Authority spokesman assured me in late February that the agency, which owned the World Trade Center until last year, had replaced many of its old lamps with newer models containing as little as 6 milligrams of mercury each. But even if you assume a mercury content lower than the 21-milligram national average, you are still facing a total release just from fluorescent bulbs of from 10 to 25 pounds of mercury that day. That does not include releases from other sources such as batteries, thermostats, and switches. All of that mercury was dispersed in minute amounts into the dust cloud on September 11. This is an enormous amount when you consider that on any given day the total nationwide average mercury emission of the several hundred coal-burning electric power plants in the United States is only 200 pounds.

Mercury is so toxic at even minute quantities that exposure levels are tightly regulated by several agencies. The EPA has a limit of 2 parts per billion in drinking water and 3.1 parts per billion in air. The Food and Drug Administration has set a maximum permissible level of 1 part per million of methylmercury in seafood. And OSHA has set a limit of 0.05 milligrams per cubic meter of metallic mercury vapor for an eight-hour shift in a workplace.

The EPA says it took a handful of tests for mercury in the air around the trade center in the first two weeks after the collapse. None detected the presence of the metal, and in a few bulk dust samples taken around the same time only trace levels of the metal were found, according to EPA spokesperson Mary Helen Cervantes. When I asked Cervantes why, with so many fluorescent bulbs inside the trade center, so few tests were taken, she replied: "We still would not consider that [the bulbs] a huge source; nonetheless, we did sample in various media."

EPA officials were equally cursory concerning their measurements of mercury in the Hudson River in the days after September 11.

On September 14, the EPA conducted tests of water draining into the river from an overflow pipe near the trade center. The test results, which were reported to the agency on September 20, showed extraordinarily high levels of mercury and other heavy metals as well as dioxins, furans, and PCBs in the water samples. For mercury, the tests showed levels of 8.8 parts per billion, four times the agency's Marine Acute Criteria, the level at which most fish are killed within an hour. In addition, tests the EPA performed the following day on Hudson River sediment opposite the trade center showed mercury levels as high as 2.8 parts per million—four times greater than the highest mercury levels measured in New York harbor back in 1993–94.

In other words, the EPA's own tests, taken within a week of the catastrophe, showed that mercury was making its way into the Hudson River from Ground Zero, even if the small number of tests the agency took for mercury showed none in the air or dust. Since then, clear evidence has emerged from other quarters of mercury contamination around Ground Zero. For instance, a private firm, that was conducting the cleanup of an office building at 90 Church Street, just northeast of the World Trade Center, has confirmed that its tests show the presence of high levels of mercury as well as asbestos and dioxin. Mercury was even found in dust behind the building's plaster walls. Cleanup was expected to take many months as work crews stripped the building down to its concrete. Any material in the building that has a porous surface has been discarded. Because the city's Legal Aid Society was headquartered in the building, cleanup crews were forced to conduct a painstaking sheet-by-sheet decontamination of thousands of important legal files.

Meanwhile, fifteen Port Authority employees assigned to work at Ground Zero were found to have elevated mercury levels in their blood. The employees, eight policemen and seven civilian workers, were immediately removed from the site. The Port Authority urged them, at the same time, to remove all fish from their diets, since high mercury levels can also result from consumption of fish that are contaminated with the metal. All the employees were later found to have returned to normal mercury levels, according to an agency spokesman.

DIOXINS AND FURANS

Dioxins and furans are the more recognizable names for two families of chemicals that have chlorine atoms attached to them. The official chemical names for the two families are Chlorinated Dibenzo-p-dioxins (CDDs) and Chlorodibenzofurans (CDFs). There are a total of 75 chemically related dioxin compounds, or congeners, and 135 in the family of furans, many of which are highly toxic. One dioxin compound, 2,3,7,8 TCDD, was identified by the EPA as the most potent carcinogen known to science. It became infamous during the 1970s as one of the components of Agent Orange, the herbicide used by the U.S. military in defoliation efforts during the Vietnam War. Dioxins are usually generated as by-products of the combustion of materials such as plastics and other chlorinated chemicals and bleached paper products, by the open burning of wood, or from the incineration of hospital and municipal waste.

The impact on humans from exposure to even low levels of dioxins and furans can be severe, since they accumulate in body fat and have a half-life of from seven to twenty years. Some dioxins not only cause cancer but can attack the liver and reproductive, immune, and gastrointestinal systems as well. In scientific tests, animals exposed to dioxins during pregnancy often have miscarriages or severe birth defects in their offspring.

Millions of pounds of plastics products were inside the World Trade Center before the collapse, including computers, telephones, plastic desks, chairs, and other furniture, and hundreds of miles of soft polyvinyl chloride cable.

"Hundreds, if not thousands, of discrete chemicals, organic compounds, heavy metals, and acids would have been produced and emitted by the collapse and uncontrolled burning of what amounted to an enormous crematorium," said Dr. Marjorie Clarke, an expert on emissions from incinerators who has testified at several legislative hearings about the trade center collapse.

Those fires, according to Clarke, were "the equivalent of dozens if not hundreds of incinerators all burning at once." But in most incinerators dioxins are destroyed when temperatures reach more than 1,800 degrees Fahrenheit. At Ground Zero the fires did not occur in an enclosed space, and each time workers removed a piece of debris more oxygen fed the underground furnace. Thus, temperatures for the Ground Zero fires were much cooler than in an incinerator. Some estimates put them at about 1,000 degrees.

"That's the perfect temperature for generating dioxins but not destroying them," Clarke said.

From the moment it began conducting the first tests for dioxins and furans in air on September 16, the EPA recorded unusually high levels. Four samples taken that day at different spots on the periphery of Ground Zero all showed the presence of several of the dioxin and furan congeners. The normal method of measuring these compounds is by calculating a TEQ (toxic equivalency) ratio that is keyed to the most potent 2,3,7,8 dioxin. The measurement is usually expressed in nanograms (one billionth of a gram) per cubic meter of air, or ng/m^3. Actual lab results on the September 16 tests, according to the EPA documents, did not come back until September 22, and they were not reported in the agency's internal daily monitoring report until September 25.

That may have been because there was one big problem: the EPA has no safety standards for many of the toxins that were found in the air at Ground Zero, including dioxin. When questioned about the dioxin standards in early March, the EPA confirmed that the New York regional headquarters had devised its own.

"Yes, we actually set benchmarks for a lot of substances where we didn't have a standard," said EPA spokesperson Mary Mears. "For many of them we didn't have standards that we can use. We did risk assessment based on what we have on these various substances." Mears said the agency's local experts in risk assessment consulted the EPA's office of research and technology in Washington and scientists at the Centers for Disease Control before they set about quickly devising "removal action guidelines" or "screening levels" based on already approved human-intake standards. In the case of dioxin, they determined thirty-year and one-year exposure guidelines. But they did so based on a cancer-risk potential of 1 in 10,000, not 1 in 1,000,000, which is the cancer-risk level that EPA policy normally seeks to achieve for a toxic substance. Under the EPA's Integrated Risk Management Information System, which was adopted in 1986, agency policy has been to reduce cancer risks to 1 in 1,000,000 wherever feasible. In some situations, the agency can allow a higher exposure risk to the public for a particular toxic substance, but it can do so only after the determination made by its staff has undergone extensive outside scientific review and after there has been a period of public comment.

Shortly after my October 26 *Daily News* article alerted the public to the elevated dioxin levels around Ground Zero, the EPA began publishing on its web page the overall TEQs for its dioxin tests. This is what the agency told the public about those standards:

> *Most of the air samples taken in areas surrounding the work zone and analyzed for dioxin have been below EPA's screening level, which is set to protect against significantly increased risks of cancer and other adverse health effects. The screening level is based on an assumption of continuous exposure for a year to an average concentration of .016 nanograms per cubic meter. Because the vast majority of individual as well the average measured dioxin levels have been lower than the screening level, EPA does not expect an increased risk of health problems as a result of dioxin being emitted from the World Trade Center site.*

The EPA's official communications to the public never mentioned that its regional office had made its own ad hoc determination of a screening level, that this standard had never existed before September 11 and had not undergone any kind of scientific peer review, and that it was based on what many scientists consider a low cancer-risk threshold of 1 in 10,000. In their defense, EPA officials were dealing with an unprecedented situation that called for rapid decisions. When they reported these "removal action guidelines" or "screening levels" as if they were some long-standing federal policy, however, they directly misled the public.

Even with those ad hoc guidelines, all four tests taken on September 16 revealed dioxin levels above Region Two's freshly minted thirty-year exposure guidelines. According to EPA records, 43 percent (32 of 73) of all dioxin tests in air the agency took between September 16 and October 18 were above its proclaimed thirty-year benchmark, and 5 percent (4) were above its one-year benchmark. Given that the thirty-year exposure level had been so hastily established at such a questionably low cancer-risk level, and without the standard review process, the high number of tests that surpassed that level should have set off alarms.

Indeed, even as the agency told the public there was no dioxin problem to worry about, its own staff was expressing a far more cautious message internally. On September 25, for instance, a daily monitoring report summarizing those first September 16 dioxin tests stated: "Levels do not pose a short-term health concern. However, elevated sufficiently to be of concern for long term [chronic exposures]. Action Item: Additional monitoring needed for dioxin beyond the debris pile perimeter."

That "action item" recommendation reflected a clear concern that dioxin was spreading beyond the "red zone" of Ground Zero into areas where civilians had been allowed to return.

Indeed, one of the few fixed monitoring stations the EPA set up beyond Ground Zero began to detect high dioxin levels from its first sample. That station, located at Broadway and Liberty Street, had been reopened to the public on September 17. But three of the first five tests at the station, taken between September 23 and October 8, found dioxin levels above the agency's thirty-year removal action guidelines.

By then, the agency had initiated more widespread and more sophisticated dioxin air sampling beyond Ground Zero, most of it conducted over a continuous twenty-four- or seventy-two-hour period.

The first such extended sample was taken between October 1 and October 4, from a sixteenth-floor window at the EPA's own headquarters at 290 Broadway, half a mile north of the World Trade Center. That sample "showed results above EPA's action level based on a 30-year exposure," according to EPA documents. The agency had now confirmed elevated dioxin levels extending half a mile away from Ground Zero, in an area where thousands of people had already been sent back to work. That should have been troubling enough. What was more troubling was that the agency did not report those findings in any of its daily summaries throughout the months of October and November. Even after it released hundreds of pages of documents under Joel Kupferman's Freedom of Information Act request, the EPA did not include the data on the dioxin tests at its own headquarters. Agency officials did take one immediate action the week they completed the testing: they quietly made respirators available to all staff at the agency that wanted them.

The dioxin tests at agency headquarters were followed by a second round of eight twenty-four-hour tests from October 11 to October 15, four of them at Borough of Manhattan Community College and four from a monitoring station on Park Row, down the street from Beekman Hospital. Both locations are several blocks away from Ground Zero. Two of the eight samples, both from Park Row, showed dioxin presence above the EPA's action level based on the thirty-year exposure. A third round of seventy-two-hour samples taken at the same two sites from October 23 to October 29 once again revealed that the two samples at Park Row had dioxin levels above the thirty-year removal action guideline. Not until a fourth round of tests in early November did the dioxin levels at Park Row finally recede. Thus it is safe to assume from the EPA's own testing that thousands of people who live or work in the Park Row area were exposed to high levels of dioxin at least from September 11 through the beginning of November, and that thousands of others who worked around the EPA building were also exposed for an unknown amount of time.

As with asbestos, the levels of dioxin in ambient air would last for only a finite time—in this case a few months. The long-term threat came from the potential for dioxin congeners to attach themselves to dust particles and penetrate inside of buildings, where they could permanently contaminate indoor spaces unless properly removed. At least three commercial office buildings near Ground Zero have since been found to be contaminated with high dioxin levels in indoor dust, while most have yet even to be tested for dioxin.

The EPA took two outdoor dust samples on October 8 from the rooftops of two buildings near Ground Zero. Both showed elevated dioxin levels, though not above the agency's removal standard in soil, which is one part per billion. (The agency chose to apply its soil standard for analyzing dust from the trade center.) The highest total dioxin TEQ on the roof of one building was 77 picograms per gram, equivalent to 77 parts per trillion, and thus substantially below the federal removal level.

That does not mean such dioxin levels are safe, however, according to David Carpenter, one of the country's top dioxin experts and a former dean of the School of Public Health at the State University of New York in Albany, who reviewed the EPA data.

"When you have 13.3 picograms per gram of TCDD [the most dangerous dioxin], as you did at that building, that is extraordinarily high for dust," Carpenter said. "Background levels in most places are about one picogram per gram. That certainly is a level of concern. It is not above the EPA standard for having to remediate, but when you have that amount in the dust samples in the presence of these huge concentrations in the air, one has to assume that the levels will accumulate in the body."

Contrary to the EPA's public assurances, Carpenter insists, "there is real reason to worry. Dioxins cause cancer and chronic disease. There does need to be concern particularly about pregnant women, or people of reproductive age who might become pregnant."

Dioxin was not just showing up in dust and air, it was also showing up in drain water and Hudson River sediment at alarming levels. In September, the agency conducted tests of runoff water draining into the Hudson and of river sediment near the trade center. "All analyzed dioxins and furans were detected," began a staff report on the September 14 sampling. The report went on to state:

> *The Toxic Equivalency (TEQ) for the sample was 122pg/L which is high. Toxic PCBs congeners were also detected at very high concentrations, with a TEQ of 151. . . . Metals and Asbestos were detected at high concentrations. . . .*
>
> *In previous harbor work performed by NYSDEC . . . the highest observed dioxin TEQ was 22 pg/L.*

In other words, EPA staff had detected dioxin levels in runoff water more than five times higher than had ever been measured in water flowing into the Hudson River. In addition, they found extremely high levels of heavy metals, asbestos, and PCBs.

On September 21, EPA administrator Whitman issued another press statement reassuring New Yorkers that their air and drinking water were safe.

"Results we have just received on drinking water quality show that not only is asbestos not detectable, but also we can not detect any bacterial contamination, PCBs or pesticides,"

Whitman reported. Technically, she was telling the truth. No contaminants had made their way into the drinking water supply, but that would have been highly unlikely anyway, since the city's water comes from hundreds of miles away in upstate New York via huge underground water tunnels. Very high levels of some potent contaminants, however, had been found in drain water coming from the World Trade Center—a clear indication that those contaminants existed in large amounts at the site—and this was almost entirely absent from her statement. Her only allusion to it was in two sentences buried deep in the press release: "Following one rainstorm with particularly high runoff we did have one isolated detection of slightly elevated levels of PCBs. This is something we are continuing to monitor very closely."

The PCB levels were not "slightly elevated," as Whitman claimed. They were astonishingly high, according to a report her own staff had produced the previous day. This is what that September 20 report said about PCBs: "Numerous PCB congeners including co-planer [dioxin-like] PCBs were detected at high concentrations. The Toxic Equivalency (TEQ) . . . is 151pg/L. In previous harbor work . . . the highest observed PCB TEQ was 0.002 pg/L."

As for asbestos levels, the EPA staff found they also were off the charts. The federal government's Maximum Contamination Level (MCL) for asbestos in drinking water is 7 million fibers per liter. The water draining from the trade center contained 9.6 billion fibers per liter, according to another EPA staff report of September 18 (see appendix). It is important to emphasize that this contaminated water was not about to get into the city's water supply. The significance of the data was as a snapshot reflecting the enormous quantity of asbestos on the ground in lower Manhattan. As with dioxins and PCBs, Whitman and other agency officials said nothing publicly about these worrisome levels. It is possible that Whitman herself had not seen the daily monitoring reports but depended on summaries from the agency's middle managers in New York City. Whether responsibility was at the top or at a lower level, however, makes no difference. The reality is that the public was not getting an accurate report of what the EPA had found.

DIESEL FUEL AND OILS

More than 130,000 gallons of oil and insulating fluid from transformers and high-power voltage lines were released at the World Trade Center on September 11 when two Con Edison substations that provided electrical power for much of lower Manhattan were destroyed by the collapse of Seven World Trade Center.

The building was erected over the substations and anchored to the ground by a series of crisscross steel beams that formed trusses over several multistory electrical transformers. Below ground level, underneath the building and the power stations, more than 40,000 gallons of diesel fuel were stored in several tanks. These tanks, which were connected by pipes to smaller tanks in upper floors of the building, provided emergency fuel to power computers for Mayor Giuliani's command center and the emergency needs of other tenants in the building. When debris from the Twin Towers set off fires at Seven World Trade, the tanks were breached. The diesel fuel inside caused such a huge blaze that the steel trusses on which the building rested weakened and gave way, thus leading to the collapse of the entire building. The collapse, in turn, ruptured the transformers and several major power lines that ran into the substations, spilling more than 100,000 gallons of oil in the transformers and 30,000 gallons of insulating fluid from the voltage lines.

In addition to the diesel fuel and oil lost at Seven World Trade Center, another 30,000 gallons of diesel were stored in separate tanks underneath other buildings in the trade center complex. While a few of the tanks were eventually recovered intact, in total 200,000 gallons of fuel and oil were lost as a result of the collapse. Since the planes the hijackers crashed into the Twin Towers each contained about 10,000 gallons of fuel, that means another twenty planeloads' worth of fuel helped feed the fires after the towers collapsed.

At least one group of scientists has found convincing evidence that petroleum was burning at Ground Zero into October. Tom Cahill, head of the UC Davis team that performed continuous independent monitoring of air quality from a spot one mile north of Ground Zero, told me in an interview of an extraordinary plume of smoke from the trade center that blew over his monitoring station on October 3.

"The wind was pushing the particles across the ground in front of our station," Cahill recalled. When he and his team analyzed the air samples, Cahill said, they found "a stunning amount of vanadium and nickel in that plume. The vanadium was fifteen or twenty times higher than any levels previously recorded in the United States." Sulfur, vanadium, and nickel in ambient air, Cahill said, are clear signs that petroleum is burning.

New Yorkers did not learn about the enormous spills of diesel and transformer oil until November 29, when I reported the information in a column in the *Daily News*. Con Edison had notified the state Department of Environmental Conservation about the spills from its properties on September 11, but had not provided details of the amount involved. After I began asking about fuel spills in early November, both Con Ed and state officials stonewalled me for several weeks. The burning of more than 100,000 gallons of transformer oils, as they no doubt understood, raised another big question—the possible release of PCBs (polychlorinated biphenyls).

For decades, PCBs were used as coolants or lubricants in transformers, capacitors, and other electrical equipment because they don't burn easily. But the manufacture of PCBs was banned in the United States in 1977 due to mounting evidence of their toxic effect. The EPA's own literature says the agency "has found clear evidence that PCBs have significant toxicity effects in animals, including effects on the immune system, the reproductive system, the nervous system and the endocrine system." By 1987 the agency had concluded that PCBs are a probable human carcinogen.

In late November, a Con Edison spokesman confirmed to me that the utility had lost 100,000 gallons of transformer oil at Seven World Trade Center. The oil, as best as the company could tell, was contaminated "with low levels of PCBs," less than 10 parts per million, according to the spokesman. New York State's definition of hazardous levels for PCBs is 50 parts per million or more. The Con Ed spokesman conceded that numerous smaller capacitors destroyed at the site could have had higher levels of PCBs. Until now, however, there have been no independent tests of oil or soil beneath the trade center by either the state DEC or the EPA that can corroborate Con Edison's statements. One DEC spokesman told me his agency was depending on the utility for test results. This despite the fact that Con Edison was fined on several occasions during the 1990s by both the DEC and the courts for falsely reporting hazardous spill information.

What is not in dispute is that unusually high levels of dioxin and furan emissions were recorded throughout October by an EPA air-monitoring station at West Broadway and Barclay Street, immediately adjacent to the destroyed Con Edison substations. The presence of furans is often associated with the burning of PCBs, according to scientist David Carpenter of the State University of New York at Albany.

The EPA's preliminary grab samples of drain water flowing into the Hudson near Ground Zero on September 14, as mentioned previously, showed extremely high levels of PCBs—75,000 times higher than any previously reported. In the first few weeks, however, the EPA told the public that this "slightly elevated" PCB reading was a onetime problem. To buttress its contention, the agency reported that several tests of PCBs in dust near Ground Zero had detected none or only trace amounts.

That report turned out to be completely wrong.

The only way to discover that error was by a close scrutiny of thousands of pages of internal agency reports. According to those documents, on the day of the trade center attack, EPA staff in lower Manhattan collected only four dust samples near Ground Zero that were subsequently tested for a variety of toxic substances, including PCBs. One sample was taken south of the site and three north of it, the farthest away being at the corner of Reade and Hudson Streets, nearly a mile to the north. Not until November 1 — six weeks after the attack — did the agency issue the results of those tests in its daily monitoring reports. All the samples had detected PCBs, the report stated, but none was higher than the EPA's standard of 1 part per million, which normally triggers cleanup efforts. Six weeks later, on December 14, the agency quietly posted a correction at the end of its full monitoring report for that day (see appendix). The correction stated:

A pesticide/PCB scan previously conducted for those four samples and presented in the Nov. 1 Sampling Situation Report incorrectly identified all levels as being below 1 ppm. Two of the samples were actually estimated to be above 1 ppm for total PCBs. The highest total PCB result of these two samples was estimated at 1.54 ppm.

This correction was astounding on two counts. First, the agency was admitting, more than three months after September 11, that high levels of PCBs had been found in the only dust samples it took the day of the attack. Second, the highest levels had been found in a sample nearly a mile away. There is now no doubt that hazardous levels of PCBs were present in the dust on some streets on the day of the attack, and it is logical to assume that similar levels could be found on nearby rooftops and windowsills, and inside residential and commercial buildings wherever windows had been left open. But by the time the test results were released hundreds of thousands of people had returned to work in the area. They had been back for nearly three months. In all probability, few building or apartment owners had bothered to test for PCBs in dust before they moved back in, since the EPA had told them there was no PCB problem to worry about.

The December 14 PCB "correction" was never announced by the EPA in a press release. Nor was it highlighted in the daily summaries the agency posted on its web page. EPA administrator Whitman issued no retraction, nor did anyone else in the agency. No one wanted to draw attention, to the fact that the agency from the start had given the public erroneous information about PCB contamination.

By including the correction at the end of its December report, however, the agency was formally complying with its responsibility to inform the public. The report in which it appeared, while technically available to the public, was accessible only through a visit to the agency's regional library at 290 Broadway. With the EPA generating new environmental monitoring reports each day, some as lengthy as fifty pages, keeping up with the latest information was a daunting task. Most reporters simply accepted whatever the EPA told them or scanned the daily summaries the agency posted on its web page. Few bothered to examine the actual raw data.

To this day very few people are aware that, contrary to official government statements at the time, the dust in portions of lower Manhattan following the catastrophe of September 11 was contaminated with PCBs. This was true not just at Ground Zero but in spots as much as a mile away.

BENZENE AND OTHER
VOLATILE ORGANIC COMPOUNDS

Benzene, a colorless, sweet-smelling liquid that evaporates very quickly, is one of the most widely used chemicals in the United States. It is found in plastics, resins, nylon, and many synthetic fibers, in gasoline, in some types of rubbers, in lubricants and dyes, in pesticides, even in some drugs, detergents, and cigarette smoke. It is extremely toxic. Breathing very high levels of benzene can cause tremors, confusion, unconsciousness, even death. Long-term exposure can cause leukemia and cancer of blood-forming organs. The chemical can also be released by volcanic eruptions or wild-fires.

For many months after the collapse of the trade center, hundreds of chemicals known as aromatic hydrocarbons and VOCs (volatile organic compounds), many of which are known carcinogens, were released into the atmosphere by the fires, and of those benzene was the toxin that appeared at extraordinarily high levels. The EPA did acknowledge from the first days of the disaster that it was finding high benzene levels at Ground Zero, and it repeatedly urged rescue workers on the site to use proper protection, as required by federal law. Unfortunately, as we shall see later, neither city officials in charge of the site nor OSHA monitored Ground Zero operations with sufficient vigor to implement that requirement.

The EPA's public announcements on benzene releases did not give any specific information or data for many weeks after the attack. Typical of its early reports was this one from the agency's Daily Environmental Summary of October 24, 2001: "Sampling for volatile organic compounds (VOCs) was conducted on October 23 in the direct area of the debris pile at ground zero. Benzene exceeded the OSHA time-weighted average permissible level at two locations."

OSHA's maximum permissible exposure limit (PEL) for benzene is 1,000 parts per billion in air for an eight-hour average, 5,000 ppb for short-term exposure. Anything greater than that requires the use of cartridge respirators or gas masks to filter out organic vapors. At two spots on the site that day, the Austin Tobin Plaza and the North Tower, the benzene grab-sample readings at the plume were 2,100 and 20,000 ppb. In fact, throughout all of September and October there were only six times when benzene readings at the North Tower plume were below OSHA's permissible exposure limit. On some days, they reached as high as 86,000 ppb (October 5) and 58,000 ppb (October 11). Even during November, readings exceeded OSHA levels in half the tests conducted, yet on some days individual samples were higher than they had been in September and October. For example, on November 8, an EPA grab sample at the North Tower plume detected 180,000 ppb of benzene — 180 times above the OSHA limit! Even as late as January 7, benzene readings were as high as 5,300 ppb.

After I reported on some of these levels on October 26, furious EPA officials insisted that they were only samples taken at ground level in the debris pile and were not an accurate reflection of the benzene hazards to workers. Benzene dissipates quickly in air, they said, and all rescue workers had been urged to wear proper respirators that would filter out such toxic gases. In an opinion piece for the *Daily News* on October 31, Christie Whitman wrote that the EPA had found "low levels of contaminants in the 'breathing zone' — 5 to 7 feet away from the debris pile — and undetectable levels away from the work site." In short, the workers had sufficient protection and the public need not worry.

Once again, Whitman was wrong. Many of the rescue workers did not have proper protection. Emmanuel Gomez, for instance, was a probationary police officer at the New York Police Department Academy on September 11. He is also a veteran reserve army lieutenant assigned to U.S. intelligence. On September 11, Gomez and his entire academy class of 800 were deployed to the World Trade Center to secure the site and help with rescue and cleanup efforts.

Six months later, during a public hearing held by EPA hazardous waste ombudsman Robert Martin, Gomez and several other police officers and firefighters testified that they had worked for weeks at Ground Zero without being provided with anything more than paper masks that were not capable of filtering out asbestos fibers, let alone benzene or other toxic vapors. Gomez, who has had extensive training in chemical and biological warfare at Fort Bragg, said that from the first day he realized the toxic threat his fellow officers were facing. Below is part of the transcript of that hearing.

GOMEZ: It was not until the third day we were issued paper masks which were not effective in any way to protecting our lungs from the harsh odors, hazardous materials and particulates in the air emanating from the fire. . . .

I noticed with my fellow officers that the superior ranking officers had masks with filtration systems. So I was seeing captains and lieutenants walking around with masks, with actual gas masks on, and here we are cleaning up the rubble—securing the area and doing everything that should be done at that time and we don't have the proper equipment.

QUESTION: For that 25 days straight, starting from Sept. 11, did anyone try to provide you a respirator?

GOMEZ: No sir.

QUESTION: Only those paper masks?

GOMEZ: Only those paper masks . . . As a matter of fact, there were only three of us in my whole company out of 50 that were actually wearing proper masks that we purchased on our own.

Later in his testimony, Gomez described the working conditions at Ground Zero: "Every time they pulled up a new slab of concrete, the flames shot up, smoke came on and a new blast of smoke and particles came out right at you."

Whitman's assurance that benzene was not a danger because rescue workers had been advised to use protective masks was meaningless. Furthermore, her claim that benzene and other pollutants did not travel beyond the actual smoke plumes on the site is contradicted by the agency's own tests. For example, on October 26, a few days before Whitman published her column in the *News*, a grab sample for benzene at Liberty and Trinity Streets, at the extreme southeast edge of Ground Zero, showed levels of 11,000 ppb—eleven times the OSHA permissible exposure limit.

CORROSIVE DUST LEVELS

An even more blatant example of how government health officials kept vital information on the dangers of the trade center's emissions to themselves is provided by events surrounding the United States Geological Survey's discovery that dust in the air over lower Manhattan was highly corrosive to human lungs.

On September 16, five days after the attack, NASA's Jet Propulsion Laboratory flew one of its reconnaissance planes over the World Trade Center site equipped with AVIRIS remote sensors. AVIRIS (Airborne Visible/Infrared Imaging Spectrometer), one of the space agency's most advanced measuring devices, is designed to detect the presence of even minute substances on the surface of a planet. It identifies minerals and chemicals by spectral analysis. Data from the spectral mapping were immediately turned over to the Geological Survey for analysis. In addition, the USGS dispatched a team of scientists to the disaster site on the evenings of September 17 and September 18 to collect dust and debris samples from thirty-five spots within a one-kilometer radius of Ground Zero.

By September 18, the USGS had provided top officials in charge of the disaster with a preliminary report and map that identified the locations of more than a dozen thermal hot spots (fires burning underground) on the site. That was followed, on September 27, by a full report from the USGS analyzing and mapping locations of asbestos and other minerals, as well as the results of chemical tests on the dust and debris, conducted both indoors and outdoors. Here, I will concentrate on what the USGS scientists discovered about the dust. Their asbestos findings will be discussed in the following chapter.

"Chemical leach tests of the dusts and airfall debris samples indicate that the dusts can be quite alkaline," the report summarized, due to high levels of concrete, gypsum, and glass fiber particles.

Anyone who has ever carried out maintenance on a swimming pool or a Jacuzzi knows that by testing for pH, you determine the acidity or alkalinity of a material. On a 15-point scale, for instance, unpolluted water will register a neutral pH level of 7. The lower the level between 7 and 0, the more acidity there is. Rainwater is often measured at pH 5 to 5.6 because of the effect of some acid pollutants. The higher the reading between 7 and 14, the more alkaline the substance is. Extremes at either end can be harmful to animals and human beings.

Most of the samples collected by the USGS team were between 9.5 and 10.5, which is around the alkalinity level of ammonia. Some were as high as 12.1—equivalent to the corrosiveness of drain cleaner.

"Indoor dust samples generated the highest pH levels (11.8) in leach tests, indicating that dusts that have not been exposed to rainfall since September 11 are substantially more alkaline than those that have been leached by rainfall," the report found.

Thus by September 27, the EPA and other health agencies in New York City had been notified by the USGS that dust in the air in lower Manhattan—both indoor and outdoor—was extremely corrosive to human lungs. Yet neither the EPA nor any other local health agency reported the findings to the public. Not until late November, when the report was published on a USGS web page, did anyone outside the government know about the findings, and even then the general public was still in the dark.

It took a February 9 investigative report by Andrew Schneider, a Pulitzer Prize–winning reporter at the *St. Louis Post-Dispatch*, to finally bring the startling alkalinity findings into the public spotlight.

"It is extremely distressing to learn that the EPA knew how caustic samples of the dust were and didn't publicize the information immediately," Joel Shufro, executive director of the New York Committee for Occupational Safety and Health, told Schneider.

EPA spokespeople repeatedly insisted to Schneider that they had warned the public about how caustic the dust was. Yet this is belied by the fact that not a single item of EPA, OSHA, or New York City Health Department literature since September 11 mentions the corrosive nature of the dust.

On February 12, the day after hundreds of people jammed a U.S. Senate subcommittee field hearing in New York to testify about the environmental problems in lower Manhattan, EPA administrator Whitman finally showed the first signs of recognizing some of the blunders her agency had committed. In a letter that day to Senator Hillary Clinton of New York, one of the conveners of the hearing, Whitman wrote: "It has become clear in recent days that despite our best efforts, some results that were developed by other federal agencies and that would have been of interest to the public were not disseminated through our web site. Even though that data was consistent with our own, I still believe the public interest would have best been served by making it available promptly."

That day, five months after the disaster, Whitman also announced that she was forming an interagency panel to gather and analyze information about indoor air quality near Ground Zero. It was her agency's first admission that there were unresolved questions about environmental safety in lower Manhattan. But even as she formed the task force she continued to repeat the agency position that "data from air quality tests thus far have been, in general, reassuring. None of the testing done to date has shown results that would indicate long-term health impacts."

As we shall see in the following chapter about asbestos, Whitman's assurances were grotesquely misplaced.

4

Ignorance, Lies, and Cover-up:
The Asbestos Fiasco

ASBESTOS IS ONE of the deadliest minerals ever used by modern industry. Estimates suggest that more than 300,000 Americans died from breathing asbestos fibers during the twentieth century. The asbestos manufacturers knew their product caused lung cancer and asbestosis but suppressed the medical evidence for decades, and bitterly fought the eventual decision of the federal government to ban many uses of the mineral while sharply restricting others. Thousands of victims whose health was crippled by working with asbestos ended up suing the manufacturers, and several firms went bankrupt from the avalanche of court cases.

The EPA and other government agencies knew that the World Trade Center contained from 400 to 1,000 tons of asbestos, primarily used as fire insulation on steel girders in the lower floors of the main towers. And they did not dispute the fact that at least some of this asbestos was present in the dust that settled over lower Manhattan. Assistant secretary of labor John L. Henshaw confirmed as much in a letter he wrote to a lawyer representing Local 78 of the Asbestos, Lead, and Hazardous Waste Laborers Union on January 31, 2002: "In that materials containing asbestos were used in the construction of the Twin Towers, the settled dust from their collapse must be presumed to contain asbestos."

The big question, though, was whether the levels of asbestos in the dust were of danger to the public. And at this point the dependability of official sources began to go seriously awry. In the six months after September 11, federal, state, and city health officials repeatedly told the public that there was no significant health risk from asbestos contamination. Officials maintained this position despite clear evidence from tests conducted both by the EPA and other government agencies as well as by private environmental firms that potentially dangerous levels of asbestos were present in large areas of downtown Manhattan.

In the following pages I will discuss precisely how the EPA, OSHA, and local city agencies misled the public about the potential hazard, how they misrepresented what constitutes federal safety standards for asbestos, how they violated federal law on proper removal of the mineral, and how each agency passed the buck and abdicated responsibility for protecting the public health, leaving thousands of tenants and property owners in lower Manhattan to deal with a massive environmental catastrophe on their own when federal help was so evidently required.

The authorities were able to get away with such negligence because the public was not able to assess the veracity of what it was being told. The specialization of environmental and product science and the complexity of government regulation relating to asbestos and other known carcinogens made it simply impossible for individual workers and residents downtown to independently gauge the risks they faced. They had little choice but to rely on what the authorities told them. And the authorities reassured them that everything was safe. EPA administrator Whitman announced on September 13, two days after the attack, that there was no asbestos problem: "EPA is greatly relieved to have learned that there appears to be no significant levels of asbestos dust in the air in New York City," Whitman said. The same EPA press statement went on to state: "Public health concerns about asbestos contamination are primarily related to long-term exposure. Short-term, low-level exposure of the type that might have been produced by the collapse of the World Trade Center buildings is unlikely to cause significant health effects."

EPA documents, however, show that the agency took only

a handful of tests for asbestos on September 11, and all of those were from Brooklyn and Jersey City, not Manhattan. Whitman's September 13 statement was thus based only on tests the agency took on September 12, when it collected its first six air samples in lower Manhattan as well as twenty-four dust samples. A few of the dust samples detected low levels of asbestos; the air samples detected none. So at that point Whitman was technically accurate, but her statement was based on limited testing by her agency across a large area.

The next day, OSHA's Henshaw issued this statement: "Our tests show that it is safe for New Yorkers to go back to work in New York's Financial District." According to Henshaw, OSHA staff wearing personal air monitors had walked through the Financial District on September 13, and subsequent analysis of the air samples they had collected showed either no asbestos at all or only low levels. In addition, Henshaw said, indoor tests of some commercial buildings had detected no asbestos, though tests of dust and debris on cars and other outside surfaces "contained small percentages of asbestos, ranging from 2.1 to 3.3—slightly above the 1 percent trigger for defining asbestos material."

On September 14, the EPA collected twelve more dust samples, all of which detected less than the 1 percent standard the federal agencies had resolved to apply at the trade center. But on the following day, of twenty-eight dust samples collected around lower Manhattan, nearly half (thirteen samples) had asbestos levels of more than 1 percent. The following day, four samples of debris collected by agency personnel at Ground Zero showed asbestos levels between 1 percent and 4 percent.

By then the agency had set up eleven permanent air-monitoring stations around Ground Zero and other parts of lower Manhattan. Each of the stations provided twelve-hour samples twice a day. The first round of samples, taken from the evening of September 16 to the following morning, detected one instance where the asbestos level was 70 asbestos structures per square millimeter (s/mm^2). The others were all below 70 s/mm^2, which the EPA would soon announce to the public as a benchmark standard for asbestos in air that it was using to determine a potential health threat. The EPA's second round of fixed air samples, taken on September 17, showed three with asbestos levels of 77, 96, and 134 s/mm^2 and another eight where, according to field staff reports, the filters of the monitoring stations had become so overloaded from all the particulates in the air that they could not be accurately analyzed.

In a September 18 statement, Whitman modified slightly her previous all-clear position: "The EPA has performed 62 dust sample analyses for the presence of asbestos and other substances. Most dust samples fall below EPA's definition of

'asbestos containing material.' " Whitman again was technically correct—most dust samples were below the 1 percent
standard; but a significant proportion were not. In a speech
to the New York State Bar Association in January 2002,
Walter Mugdan, chief counsel for the EPA's New York region, reported that "around 35 percent of the samples of
bulk dust taken in lower Manhattan in the first few days after
the collapse exceeded the 1 percent value."

That was a crucial admission, for it indicated that a substantial portion of the dust samples taken over several square
miles of lower Manhattan streets, from Battery Park City to
Canal Street, were asbestos contaminated. If this was the
case for samples on the street, then dust on rooftops and window ledges, as well as that which had made its way into
homes and offices, must have been similarly contaminated.

While it was significant that the EPA and OSHA repeatedly played down the fact that their tests were regularly showing levels of asbestos above the standards they deemed safe,
on another level these test results were meaningless because
they employed standards of safety that were completely un-
reliable. This was a point repeatedly emphasized by Cate
Jenkins, a twenty-two-year veteran scientist in the EPA's hazardous waste division, and by the agency's ombudsman
Robert Martin. Both have since charged that OSHA and the
EPA were misrepresenting federal law and their own regula-

tions in their use of the 1 percent and the 70 structures per square millimeter standards. As Jenkins and Martin point out, the EPA has never adopted a safety standard for asbestos in open air and neither of these measures when first introduced was intended as such.

Let's look at these two standards, the 70 structures per square millimeter standard first. This was developed not as a general standard for safe levels of asbestos but as a test for the successful removal of asbestos from public school premises. Before children are permitted to return to a school where asbestos has been professionally removed, air samples are taken in individual rooms using a method called "aggressive testing." This means the abatement firm in charge of the removal must operate a one-horsepower leaf blower in the closed room for at least an hour before the samples are taken to assure that as many fibers as possible become airborne. The school can be reopened only if a subsequent air sample passes the 70-structure test.

The 70-structure level was arrived at because it was the detection limit of the testing filters that were in use when the regulations were adopted in 1986; in other words, it was simply not possible at the time to measure the presence of asbestos at lower levels. But it is not a safe level—the EPA's starting point for considering remediation for any toxin is known as the one-in-a-million threshold: if more than one person in a million could develop cancer from exposure to a substance, then the EPA must review the risk and consider appropriate action. The 70-structure level is two thousand to four thousand times higher than the EPA's one-in-a-million cancer-risk level, according to Jenkins. It is, for example, four to eight times higher than asbestos levels inside homes in Libby, Montana, which the EPA had already declared a Superfund site.

Consider now the 1 percent standard, the other test level that the EPA employed to measure whether dust in the air downtown was safe. This standard is described in the EPA's own literature in the following manner: "In April 1973, the EPA issued the National Emission Standards for Hazardous Air Pollutants (NESHAP) for asbestos. The NESHAP regulation governs the removal, demolition and disposal of asbestos-containing bulk waste. An asbestos-containing product, as stated by the regulation, was defined for the first time to be a product with greater than 1% asbestos by weight. . . . It must be understood that the EPA NESHAP definition of 1% by weight was not established to be a health-based standard."

In other words, the 1 percent standard has nothing to do with what is a safe level of asbestos in air or dust. It is rather a standard by which the federal government determines whether any construction material has enough asbestos content to require professional abatement. Clearly, the dust around the World Trade Center site was not a building product in its own right. But the asbestos in that dust, at whatever level, had originated from the collapse of the Twin Towers, from building materials that contained far more than 1 percent asbestos. In fact, some samples of coatings on steel beams at the trade center have revealed asbestos content as high as 40 percent.

If the 1 percent standard had no relation to safety levels, why did the EPA and OSHA rely on it from the beginning? According to Jenkins, they used it because the testing method adopted for most of their dust samples, a method called Polarized Light Microscopy (PLM), cannot detect asbestos levels lower than 1 percent. The EPA's Region Two chief counsel Walter Mugdan admitted as much during his speech to the New York State Bar Association, when he said, "The 1 percent standard is not necessarily health- or risk-based, but rather keyed to the detection limits of the specified analytical method." By using this older and cruder test method, however, the EPA was violating its own regulations, which require that all samples of settled dust or other solids be tested for asbestos using a more up-to-date method called Transmission Electron Microscopy (TEM), capable of detecting at lower levels. Surprisingly, the top EPA officials from Region Two had been offered additional TEM equipment by another of the agency's regional offices within the first few days of the disaster but turned down the offer.

The simple reality is that the U.S. Congress, when it adopted the Asbestos School Hazard Detection and Control Act in 1986, made it clear that there is no safe level of exposure to asbestos. In the statement of findings and purposes of that law, Congress stated: "Medical science has not established any minimum level of exposure to asbestos fibers which is considered to be safe to individuals exposed to the fibers." Nor is it true that cancer can be contracted from asbestos only after many years of exposure, as Whitman and the EPA initially told the public. While prolonged exposure certainly increases the likelihood of contracting the disease, there are documented examples of people who were subjected to short-term high exposure to asbestos fibers and developed cancer many years later as a result. The Agency for Toxic Substances and Disease Registry, which as part of the Centers for Disease Control is responsible for developing toxicological profiles for chemicals and other toxins, has determined that the "cumulative exposure" or total dose method is valid for predicting how asbestos exposure may affect human health. The key factor is how many fibers get lodged in the lung and are able to do damage over a long period, not simply how long a person is exposed to the fibers.

The EPA was aware that levels of asbestos far lower than

1 percent in dust or soil can be dangerous. Jenkins, in an extensive and well-documented March 11 memo that blasted her agency's handling of the crisis, cited public guidelines that the EPA's Region Eight had issued during its cleanup action at an asbestos-contaminated Superfund site in the town of Libby, Montana.

QUESTION: I recently read that EPA found less than 1% (or trace levels) asbestos at Fireman's Park and other locations that were sampled. Is that a safe level?

ANSWER: This is a very difficult question, and at this time we are not sure. Levels of 1% or less may be safe. Even higher levels could be considered safe at remote locations where no one comes in contact with the material. The key to determining whether there is risk is exposure. If there is no exposure pathway, i.e., a way for the asbestos to get into your body, such as contact with the material, or people driving over the material so that they breathe in the fibers, there is no risk. Levels of 1% or less could present a risk where there is enough activity to stir up soil and cause asbestos fibers to become airborne.

Of course, at the World Trade Center the asbestos was not present in soil but in dust. And dust, obviously, is more easily airborne. The extent of its dispersion across lower Manhattan was made clear on September 27, when the United States Geological Survey forwarded to federal and city health officials the results of its satellite mapping of the area. One map showed hundreds of tiny red pixels indicating the asbestos hot spots throughout lower Manhattan. Each of those pixels, one of the scientists involved in the study told me, represented an area of six by twelve feet either on the surface of a street or on the roof of a building where NASA had detected chrysotile, the type of asbestos most commonly found at the World Trade Center site. The USGS team reported that chrysotile was detected in a little more than two-thirds of the thirty-three dust samples it took. The report suggested a clear pattern in the dispersal of the asbestos dust: "It is apparent that trace levels of chrysotile were distributed with the dust radially in west, north, and easterly directions, perhaps at distances greater than 3/4 kilometer from ground zero." In the debris from one steel beam at Ground Zero, the team detected a chrysotile asbestos content of between 10 percent and 20 percent.

The USGS report provided a clear map for the EPA and city health officials of where the greatest potential asbestos contamination existed. It made clear that the dispersal of the dust had been widespread but uneven. None of this information was made available to the public by the various government health agencies despite the fact that the pathways of exposure to humans from asbestos in city dust are far more extensive than they would be in a rural setting like Libby, Montana. Dust was particularly effective in carrying the asbestos indoors, through open or ill-fitting windows and doors.

This brings us to the critical issue of how the authorities handled the matter of indoor air. Jenkins makes the point that tests of outdoor air are irrelevant to determining the concentration of asbestos in indoor air. Outdoor air, after all, is easily diluted by changes in wind currents or by rainfall. Indoors, however, asbestos fibers have no place to go. They are simply recirculated by air-conditioning and heating systems. Any fibers in settled dust will eventually become airborne as a result of human activity—sweeping or dusting, even walking will send the contaminant back into circulation. In this way, asbestos levels can remain high for years in buildings that are not professionally abated.

The EPA restricted its responsibilities to monitoring outdoor air, while the Giuliani administration assumed primary responsibility for monitoring and cleanup of the insides of commercial and residential buildings. The EPA's web site referred New Yorkers to the city's Department of Health and its Department of Environmental Protection for guidelines on the cleanup of buildings. The Department of Health issued recommendations for cleaning homes and office space on September 17 (see appendix). These included the following:

The best way to remove dust is to use a wet rag or wet mop. Sweeping with a dry broom is not recommended because it can make dust airborne again. Where dust is thick, directly wet the dust with water, and remove it in layers with wet rags and mops. Dirty rags can be rinsed under running water (try not to leave dust in the sink to dry). Used rags and mops should be put in plastic bags while they are still wet and bags should be sealed and discarded. Cloth rags should be washed separately from other laundry. Wash heavily-soiled or dusty clothing or linens twice. Remove lint from washing machines and filters in the dryers with each laundry load. Rags should not be allowed to dry out before bagging and disposal or washing. . . .

If your apartment is very dusty, you should wash or HEPA vacuum your curtains. If curtains need to be taken down, take them down slowly to prevent making dust in the air. To clean plants, rinse leaves with water. Pets may be washed with running water from a hose or faucet; their paws should be wiped to avoid tracking dust inside the home.

These recommendations were in clear violation of federal safety laws. The dust produced by the collapsed buildings, as John Henshaw clearly accepted in his January 2002 letter, contained asbestos. Since that dust had been emitted from materials containing more than 1 percent asbestos, all of it had to be considered hazardous, under the NESHAP regulations of the federal Clean Air Act, and could be removed only by certified asbestos removal contractors. Those regulations require that a contractor file specific plans for abatement with the local agency that monitors asbestos cleanup.

The New York City Charter names the DEP as the agency charged to "respond to emergencies caused by releases or threatened releases of hazardous substances" and also "to collect and manage information concerning the amount, location and nature of hazardous substances." Under those provisions of the charter, the DEP had assumed responsibility over the years for the regulation of any asbestos removal projects, though this was a tiny part of its functions. While the agency did urge building owners to test the dust inside their buildings for asbestos, the instructions it provided were as erroneous as those that the city's Health Department put forward to the general pubic.

"EPA is using the 1% definition in evaluating exterior dust samples in the Lower Manhattan area near the World Trade

Center," DEP commissioner Joseph Miele wrote in an October 25 letter to lower Manhattan residents. "All affected landlords have been instructed to test dust samples within their buildings utilizing this standard. Landlords were notified that they should not reopen any building until a competent professional had properly inspected their premises. If more than 1% asbestos was found and testing and cleaning was necessary, it had to be performed by certified personnel."

The problem with Miele's message, once again, was that the Clean Air Act's 1 percent standard is not for dust but for the material from which the dust originates. All dust from the World Trade Center should have been considered asbestos contaminated, even at levels lower than 1 percent. Furthermore, Miele's agency failed to establish any procedures for building owners to report contamination in their buildings to the city. Under normal conditions, an owner who discovered asbestos contamination would have to file plans with the DEP on how he intended to remove the material. Given the unprecedented emergency, the DEP waived all such requirements and conducted little if any monitoring of landlord compliance with asbestos abatement laws. During a U.S. Senate subcommittee hearing in February 2002, Miele testified that his department had only sixteen employees assigned to asbestos oversight and that his agency had been overwhelmed just by the requirements necessary to ensure that the city's water supply was safe.

The city's inability to oversee the cleanup created a gaping hole for unscrupulous landlords to perform slipshod abatement, or to do nothing at all. Some owners did perform responsible testing and cleanup, but horror stories abound from downtown residents about shoddy efforts. Eric Mandelbaum, for example, has lived for decades in a luxury apartment building on Gold Street, several blocks east of the trade center. A retired city worker, he is also president of his building's tenant association. He and several other tenants were furious when they returned to their building the week after the attack and discovered that one of their janitors, together with a couple of immigrant day laborers, had been dispatched to use household vacuum cleaners to remove the thick dust that was coating their apartments. "The building management told us that the tenants would have to be there to move our own furniture while these men cleaned," Mandelbaum said.

Marisa Ramirez de Arellano confronted a similar harrowing experience with her ninth-floor apartment at 333 Rector Place in Battery Park City. The day of the attack, large amounts of dust from the collapsed Twin Towers entered the apartment through an open glass sliding door. With the electricity gone, she evacuated for several weeks. The building's

management took air samples for asbestos—no other contaminants—and reported to her that no elevated levels were
found. By the end of September she had developed breathing problems. "As soon as I returned to the apartment, my
breathing problems worsened," she said. "The apartment
had been 'cleaned' but there was still gray dust everywhere."
When she complained to the building management they
agreed to send a professional cleaning firm to do the job
again. "They brought brooms and a vacuum cleaner; none
of them were wearing dust masks or respirators as I would
have expected a hazardous material cleaning crew to wear,"
she wrote in an official complaint to the city's Department of
Environmental Protection. The workers who arrived were
all immigrants who spoke only Spanish and "had no idea
how to clean a contaminated room," Ramirez stated. After
the second cleanup, so much dust still remained that she
and other tenants arranged for their own testing of the dust.
A firm they hired reported that the dust contained up to
2 percent asbestos.

Another tenant of a luxury rental building on Greenwich
Street, south of Ground Zero, said his landlord refused to
make available the results of air-monitoring tests for the
building and also used immigrant day laborers for the
cleanup. "Every time someone sits on my living room couch
a puff of dust shoots up in the air," he said.

When a group of local officials attempted to get a clearer sense of the potential indoor contamination, they were rebuffed by city authorities. On September 15, a dozen politicians representing lower Manhattan formed themselves into a Ground Zero Task Force. Kathryn Freed, who was the area's City Councilwoman at the time, asked officials in the Giuliani administration if the group could conduct some independent environmental testing of a few buildings in her district. City Hall denied the request. On September 18 Freed resorted to sneaking a team of environmental scientists past police barricades so they could secure air and dust samples in two residential buildings within the restricted zone around Ground Zero.

Those scientists, from Cincinnati-based Environmental Quality Management Inc., reported astounding levels of asbestos. In the first building, four blocks north of Ground Zero, where only a thin layer of dust had found its way indoors, the group found from 279 to 376 fibers per square millimeter—four to six times higher than the 70-fiber EPA standard. In the second, a "high dust" building three blocks southwest of the trade center, the team detected from 6,200 to 10,600 fibers per square millimeter—up to 150 times the EPA standard. Even these extraordinary levels understated the contamination, since they were conducted under "passive" testing methods, that is, without using a fan or blower to whip up the dust as is required for the 70-fiber standard. The implications of the Ground Zero Task Force tests were frightening. If two randomly chosen residential buildings near Ground Zero had such high levels of asbestos, it was clear that many other buildings in the area must have similar contamination.

This fact was underscored on October 9, when I reported in a *Daily News* column that another independent firm, Virginia-based H.P. Environmental, had found similar startling results from the testing of two commercial buildings near Ground Zero. "In the beginning we were getting clean results, or no [asbestos]," said Piotr Chmielinski, one of the scientists who worked on the study. "Until we discovered that a lot of dust and other particulates in our samples were obscuring the tiny asbestos fibers in the background." Many indoor tests being done by federal or local health officials or by private cleanup companies used a testing procedure called phase contrast microscopy, which detects only larger asbestos fibers of at least 5 microns in size. But when Chmielinski and the team leader, Hugh Granger, used a more exact—and more expensive—method of analysis, electron microscopy, the same samples they had tested with the cheaper method showed far greater amounts of asbestos.

The team took eleven air samples on different floors of two buildings that were up to three blocks from the trade center site. Using the cheaper and less exact method, only two of those samples revealed asbestos levels higher than federal permissible exposure limits. But when they examined those same samples under the electron microscopy procedure that detected smaller fibers, they found that seven of the eleven samples had asbestos levels above the EPA's 70-fiber standard. Three samples had more than 300 fibers and one registered 369 fibers.

That is a "magnitude of fibers that it would be best not to ignore," Chmielinski warned. According to Chmielinski and Granger, an unusually high percentage of dust resulting from the trade center disaster was composed of these smaller microscopic fibers, most likely because of the high level of pulverization caused by the enormous pressures produced in the collapse of the Twin Towers. While some scientists believe that smaller asbestos fibers are unlikely to do as much damage to lungs as large fibers, and thus do not pose the same danger, others disagree. "The smaller the particle . . . the easier job that it has penetrating right down into the very depths of the lungs," contends Dr. Philip Landrigan, director of environmental and occupational medicine at New York's Mount Sinai Medical Center.

H.P. Environmental's discovery that asbestos contamination in buildings near Ground Zero was greater than government officials were claiming made a lot of people nervous.

After all, this was the same firm that the Port Authority of New York had used as its environmental consultant after the first World Trade Center terrorist bombing in 1993, so its conclusions could not easily be dismissed. Instead, like so much troubling information about the disaster, attempts were made to bury it.

In early October, Chmielinski joined Granger in a meeting with several top contractors and city officials involved with Ground Zero cleanup. "We walk into a huge conference room with, like, thirty people, and those guys who took us in are sweating," Chmielinski later recalled. "Hugh and I were joking. What are we worried about? We have our data, we do our job." They received very little visible reaction to their report from the industry and government insiders who were present that day. Shortly afterward, however, Chmielinski and Granger arranged to post their findings on the web site of the American Industrial Hygiene Association in order to make the information available to the media and the general public. Within five hours of the posting, their report had been removed from the site without any explanation. The next day Chmielinski and Granger were told that their services were no longer needed at Ground Zero. A copy of their report eventually reached my desk and I wrote about it in the *Daily News*.

Thus, by mid-October testing conducted by both the Ground Zero Task Force and H.P. Environmental had made it clear that there was significant asbestos contamination in the indoor air of buildings at least within a radius of a quarter mile of Ground Zero. New York City's health agencies had already demonstrated their ignorance about the basic requirements of federal law when it came to asbestos contamination, and they were well aware that they did not have the personnel to monitor cleanup activities of building owners.

If ever a public health crisis called for federal intervention, this was it. Yet the EPA continued to insist that it did not have legal responsibility to monitor or clean up the insides of privately owned buildings. This was another misrepresentation, according to both U.S. Representative Jerrold Nadler and the EPA's own ombudsman. The EPA had full authority from the beginning of the crisis to take over responsibility for the monitoring and environmental cleanup of both indoor and outdoor spaces under the federal government's National Contingency Plan for dealing with national emergencies, which is authorized by the Comprehensive Environmental Response, Compensation and Liability Act. This act stipulates that the EPA is the government's lead agency for handling the release of hazardous substances that might present an imminent and substantial danger to public health. Moreover, a presidential directive signed by Bill Clinton in 1998 makes the EPA's responsibility for the indoor cleanup of lower Manhattan even clearer.

EPA administrator Whitman admitted as much herself during testimony before a U.S. Senate appropriations sub-committee on November 28, 2001, when she said, "Under the provisions of PDD 62, signed by President Clinton in 1998, the EPA is assigned lead responsibility for cleaning up buildings and other sites contaminated by chemical or bio-logical agents as a result of terrorism. This responsibility draws on our decades of experience in cleaning up sites con-taminated by toxins through prior practices or accidents."

That is why EPA ombudsman Robert Martin, in a letter summarizing the preliminary findings of his investigation into the agency's World Trade Center response, concluded that "EPA has not fully discharged its duties under PDD (Presidential Directive) 62, the National Contingency Plan (NCP), and the 2001 OMB Annual Report to Congress on Combating Terrorism. EPA has abandoned its responsibili-ties for cleaning up buildings (both inside and out) that are contaminated, or that are being re-contaminated, as a result of the uncontrolled chemical releases from the World Trade Center terrorist attack."

Ironically, on September 7, 2001, only four days before the World Trade Center attack, Whitman and the EPA assumed total responsibility for just such a total cleanup in the town of Libby, Montana, a former center for asbestos mining where, as mentioned previously, asbestos levels in the soil and dust were lower than those subsequently found in lower Manhattan.

"It has never been our plan to look to you to pay for any part of this cleanup, including the cleanup of residential properties," Whitman told the residents of Libby.

What most likely is preventing the federal government from treating New York City as it did Libby is the enormous difference in cost. There are only a few thousand residents in Libby, and only a few hundred homes that must be cleaned. A full environmental cleanup of lower Manhattan, however, would run into hundreds of millions of dollars. Nonetheless, in the same week that EPA hazardous waste ombudsman Martin issued his preliminary report, New York City's new mayor, Michael Bloomberg, announced that the city, with the help of funds from the EPA, would begin a massive cleanup of the outside facades and roofs of some 200 buildings in lower Manhattan to prevent any asbestos or other toxins that had lodged on building surfaces from recontaminating the area.

That the city and the EPA were starting such a cleanup more than six months after the disaster was itself a tacit recognition of their past failures, for if the cleanup was indeed necessary it should have been undertaken at the very beginning, before the public was given the go-ahead to return to the area. On one matter, however, government safety officials still refuse to budge; they continue to resist responsibility for the testing and cleanup of the insides of all buildings.

U.S. Representative Nadler, who has led the fight in Congress to expose the agency's failures, insists that "the EPA has no legal basis to shirk its responsibility" for a "total cleanup." When it comes to preventing illness or premature deaths from contaminants released by the trade center collapse, Nadler says, the federal government must assume the cost.

5

The Rescue Workers: Abandoned Heroes

IN THE FIRST WEEK after the attacks, thousands of firefighters, police, emergency rescue workers, even private citizens rushed to Ground Zero—many of them from as far away as California—to help in the monumental work of digging out victims, fighting the fires, and removing debris. Their bravery and their willingness to risk their own safety inspired all of America. Hundreds of government officials who supervised the complex recovery under enormous pressures likewise earned the public's admiration and gratitude. But my own interviews with scores of these workers, along with a handful of preliminary surveys and public hearings that have looked into the recovery effort, reveal a dark side of the Ground Zero saga. Top city and federal officials failed to enforce even the most basic health and safety procedures at the World Trade Center site for weeks and even months, thus abandoning those heroic workers to dangerous toxic emissions that are likely to cause unnecessary disease and death in the years to come.

An October 6 report by investigators from the National Institute of Environmental Health Sciences, for instance, revealed the huge problem faced by the more than 5,000 emergency personnel at the site during the first few weeks.

"No WTC disaster project safety and health plan apparently exists as of October 5, 2001," NIEHS investigators John Moran and Donald Elisburg wrote. The workers, the study found, "are tired, weary and extremely fatigued and they are operating in an environment essentially devoid of any organized safety and health protection programs." Many of the rescue personnel were not wearing proper protection against the enormous quantities of toxic substances known to be on the site, the report noted, adding that of the 1,350 construction workers at Ground Zero, "probably fewer than 20% have been medically certified to wear respiratory protection or have had respirator training."

A major roadblock to implementing normal safety procedures, Moran and Elisburg reported, was that nearly a month after the attacks the site was still officially being designated as an emergency rescue operation.

"While the New York City Fire Department was 'in charge' of the WTC Disaster site," Moran and Elisburg wrote, "the Incident Commander was, as a practical matter, the Mayor through the New York City Mayor's Emergency Management structure." This centralized command structure at the site was presenting a "serious challenge" to handling health and safety issues, they noted.

While the Fire Department was nominally in charge of the site, the reality was that the department was headless — its most experienced commanders had been killed on September 11.

"There was nobody in charge," said Rudy Sanfilippo, a leader of the Uniformed Firefighters Association, who spent much time at Ground Zero and who insists that operations there were "out of control" for weeks.

"The chief was dead. The deputy chief was dead. Ray Downey, the chief of operations, was dead," Sanfilippo said, while Fire Commissioner Thomas von Essen played almost no visible role at the disaster site itself.

"He [von Essen] rode off into the sunset with Giuliani, holding press conferences every day, rising to fame on our backs," Sanfilippo claimed bitterly, "and he absolutely abandoned the men."

Several police officers and firefighters, as cited earlier, testified at public hearings held by the EPA's ombudsman that they were not given proper respirators for weeks.

A week after the attack, Sanfilippo claims, he warned department officials that they needed to get respirators for all firefighters on the site. "They gave us their word they would," he said. "It never happened. It was months before they finally got the proper filters to attach to our self-contained breathing masks." Gas masks with oxygen tanks that the department holds in stock were useless at Ground Zero because they have sufficient oxygen for only two hours.

David Prezant, deputy chief medical officer for the New York City Fire Department, confirmed in a December 10 speech at a conference of the National Institute of Occupational Safety and Health that some improper equipment "of questionable value" had been issued to firefighters and that a disturbing number of them wore no respiratory protection. During the first week of the tragedy, Prezant said, 70 percent of firefighters he surveyed used a simple dust mask. By the second week, 58 percent were using respiratory protection either rarely or not at all. Some of the problems, Prezant conceded, were due "to confusion and delay" in supplying the right respirators.

By early March, the Fire Department had 200 firefighters on medical leave and nearly 700 had reported some sort of respiratory problems from World Trade Center duty.

"What's disgusting," says Sanfilippo, "is that six months later everyone is just avoiding the mistakes they made. We look at these government agencies — is anyone telling us the truth? This was not handled as a hazardous material incident from the beginning, and now our men are paying for it. There was nothing in the dust? So where did it go? We're not fools."

Firefighters are not the only victims of the city's bungling. According to a report from the Natural Resources Defense Council, "OSHA inspectors reportedly observed dozens of workplace safety violations on a daily basis in late September and early October at Ground Zero, but did not take action to insure that proper respirators were worn."

Kevin Mount, for example, has been a heavy-equipment operator for the city's Sanitation Department for twenty-two years. On the night of September 12 he was ordered to Ground Zero to drive dump trucks filled with debris away from the site. After two weeks he was transferred to the city's huge trash landfill at Fresh Kills in Staten Island, where he normally worked. He was assigned to operate a bulldozer in a restricted area where debris from Seven World Trade Center was being dumped, then systematically examined by law enforcement agents. The conditions for some workers at the landfill were far worse than those at Ground Zero, where the constant eyes of reporters and government officials assured at least some minimal attention to safety.

Mount and scores of other sanitation workers toiled twelve hours a day for seventy-five days until they got their first day off for Thanksgiving. For the initial forty-five days the only safety equipment they received was paper dust masks. Sometimes the dust around them was so thick that Mount could barely see, because the FBI, to preserve evidence, had banned the hosing down of the debris with water. The faster that Mount and the other men worked, the more dust their machinery kicked up.

"He was feeling sick so many times throughout that period," his wife, Heidi, told me. "But Kevin knew there was a job to be done. It was a matter of loyalty and patriotism to him. When he came home loaded with dust I treated him like a hero. He'd say, 'Don't touch me, 'cause I don't know what's on me.'"

Mount and the other sanitation workers noticed that the FBI agents and police detectives searching for evidence in the mounds of debris were equipped with Hazardous Materials suits, including goggles, respirators, and rubber boots. So were the people from the Federal Emergency Management Administration and even the OSHA inspectors on the scene. They saw wash stations erected where those agents and officials were decontaminated each day. The sanitation workers complained daily to their supervisors and union officials and demanded protective gear, but each time the word came back that the air was fine, that even the mayor had said it was safe.

By early February, Mount had become too ill with respiratory problems to work. A visit to his doctor revealed he had an acute case of hepatitis C and abnormalities in his liver, while a chest X ray showed he'd experienced significant lung changes. On February 19, suffering from a high fever, gasping for air, and with a constant burning sensation in his tracheal area, Mount was hospitalized for several days. Doctors administered steroids and antibiotics and prescribed three different inhalers for him. Once he was released from the hospital, a new battery of blood tests revealed he had elevated levels of manganese in his body.

Six months after the collapse of the Twin Towers, Mount, forty-six, a man who rarely missed a day's work in twenty-two years as a city employee, had become a very sick man.

"His breathing is deteriorating each day," his wife said. "He can't walk up a flight of stairs without being out of breath."

The Mounts and scores of other Ground Zero workers are furious at the way some of them were abandoned by the city, and how OSHA and other federal agencies failed to protect them. "They rode us hard and put us away wet," Kevin Mount says.

In addition to the rescue and city workers, hundreds of

iron workers and operating engineers who worked at Ground Zero for private contracting companies are already experiencing health problems.

Dr. Stephen Levin, of Mount Sinai's Selikoff Center for Occupational and Environmental Medicine, has been conducting medical screening of many of those workers.

"A high percentage of them are still experiencing respiratory problems six months later," Levin said. And a "very high percentage" have been unable to cope with the emotional trauma of their experience.

"Time and again, I've watched these men break down in tears as they begin to talk about what they went through," Levin told me. Virtually no one is paying attention, he believes, to the significant post-traumatic stress from which so many of the recovery workers are suffering.

The emerging health crisis with rescue and recovery workers has already spread beyond New York. As of February 11, 2002, out of seventy-four emergency responders from Ohio Task Force One who had volunteered to work at Ground Zero, thirty-seven had become ill, according to Robert Hessinger, the logistics chief for the task force. Three were hospitalized with viral pneumonia, eight experienced extreme weight loss, two have been diagnosed with the onset of adult asthma, and one has come down with acute bronchitis; the rest have experienced respiratory disorders and rashes.

Likewise, in California, of 395 emergency responders who volunteered to work at Ground Zero for the first three weeks after September 11, 100 have already filed workers' compensation claims because of illnesses.

"I am concerned, and I remain concerned, that no Federal agency is monitoring these workers for health problems," U.S. Senator George Voinovich from Ohio said in a written statement to a senate subcommittee in February.

According to the study of Ground Zero conducted by the National Institute of Environmental Health Sciences, part of the reason worker safety laws were not enforced during the first weeks after September 11 was that all action on site at first was classified as an emergency rescue operation and thus all safety laws had been suspended "until such a time as the Incident Commander terminates the rescue phase and turns the site over to cleanup operations."

In other words, by continuing to classify the Ground Zero operation as an emergency rescue effort, Mayor Giuliani kept complete operational control of the site in his hands. And because of that, the city was then able to ignore federal and state laws regulating proper health and safety procedures. Given the constant media accolades Giuliani was receiving for his management of the crisis, officials from OSHA, the EPA, and state health agencies proved powerless, or at least unwilling, to publicly confront the mayor over the city's lack of compliance with federal safety laws at Ground Zero—just as they failed to confront the Giuliani administration's failure to properly monitor indoor air quality in the rest of lower Manhattan.

Both failures are certain to haunt New York City for decades to come. Already more than 1,300 notices of claims totaling more than $7.2 billion have been filed in the state court system, many of them by rescue workers, alleging that the city mishandled health and safety laws at the World Trade Center disaster site.

It will be many years before we get an accurate picture of the long-term health impact on Ground Zero workers resulting from their exposure to so many toxic substances. By then most of the government leaders responsible will be long retired, but it is likely that the damage will be staggering.

6

Uncovering the Truth

SHIRLEY KWAN is typical of the thousands of New Yorkers who returned to lower Manhattan after September 11 and could not reconcile what the government kept telling her with how she was feeling. Kwan, twenty-eight, lives in a small apartment on Mott Street in Chinatown, about a mile north of Ground Zero, and each day she walks to work in the financial district. Her parents, who have lived in Chinatown for decades, have their own apartment a few blocks away. Theirs is the poorest neighborhood in downtown Manhattan and it has received perhaps the least media attention of any of the affected areas, though its economy was devastated in the weeks following the attack.

"There was basically nothing done in Chinatown," Kwan told me in December. "None of the buildings were cleaned by the landlords. There were no community forums, the government never came to explain anything to us."

Kwan kept hearing the EPA and city Health Department reports stating that the air was safe, yet every night the acrid air seeping into her and her neighbors' apartments was making them all sick.

"It gets so bad you can't even sleep," she said. "It's a burning smell and it stays in your apartment. I had two months of bronchitis. I was coughing so badly that all the muscles on my rib cage hurt. On one occasion I had to go to the emergency room, so my doctor suggested I get out of there."

Kwan moved in temporarily with a friend in Queens, and the respiratory problems began to subside. But her parents were still stuck in their apartment. Immediately after the attack, she began going door to door in her neighborhood trying to assist the Chinese immigrants city officials were ignoring.

"The second day of the tragedy I asked the Red Cross for respiratory masks for my neighbors," she said. "They said, go to the Health Department. They were closed up. We called city hospitals and asked for donations and distributed them.

"Many of the elderly had asthma," she said. "A lot of people had just lost their jobs in restaurants and the garment industry, so health for them became secondary to having a job.

But health is a treasure, I told them. If you have no health, you can't do anything else."

Kwan eventually joined with Maureen Silverman and community leaders from other downtown neighborhoods to form the World Trade Center Emergency Environmental Group.

Much like Kwan, Marilena Christodoulou was shocked as she gradually realized that health officials both in the city government and at the Board of Education were lying to her about the environment around Ground Zero. Christodoulou is the president of the Stuyvesant High School Parents Association. The school, only a few blocks north of Ground Zero, was closed for nearly a month after September 11. During that time, its 3,000 students were transferred to Brooklyn Technical High School, another school for the academically gifted. The Stuyvesant parents, many of them from the city's business elite, pressured the Board of Education to rapidly clean up any environmental problems at the school. The school board spent nearly $1 million on the cleanup and reopened Stuyvesant for classes on October 9.

Once the students were back in the building, however, Christodoulou and the other parent leaders found that many of their children as well as some teachers were becoming ill with respiratory problems. They learned, in addition, that the city had set up an operation at a pier a hundred feet from the school to load debris from the trade center site onto barges and ship it to the landfill in Fresh Kills. Hundreds of trucks a day lined up in front of the barges and the school to unload the debris, unleashing constant clouds of dust. What was even worse, Christodoulou was surprised to discover that the Board of Education had not cleaned the school's air-conditioning system nor upgraded the system's filters to block pollution from coming into the school, as the parents had been promised. The parents' association then hired its own environmental consultant, and the tests he conducted showed that on many days levels of particulate matter inside the school were above EPA safety limits.

As the weeks passed, and the parents demanded even more extensive air monitoring, new tests revealed further disturbing information. As late as February, for instance, levels of lead dust in many Stuyvesant High classrooms were far above federal safety limits. Christodoulou rapidly emerged as one of the most outspoken critics of the web of environmental deception that public officials were weaving.

Perhaps the key figure in the search for the truth in this environmental fallout tragedy is Joel Kupferman, the unpretentious lawyer who has run the tiny nonprofit New York Environmental Law and Justice Project on a shoestring budget for years. In interviews with reporters, Kupferman tends to

speak in rapid-fire sentences that jump from one environmental issue to another. But even before September 11 he
had earned respect among federal and state officials as a bull-
dog advocate for public health. His second-floor loft office on
lower Broadway, just across the street from the EPA headquarters, soon became the center of community resistance to
the "official story" that the air around Ground Zero was safe.
As Kupferman walked the streets during those first few days,
everyone from firefighters to cops to office workers would
stop him to complain about their respiratory problems.

By late September, Kupferman had filed Freedom of Information Act requests with every agency involved in health
and safety. Had it not been for his pressure, and for the press
attention generated by the documents he secured, it is unlikely that much of the data about environmental contamination from the World Trade Center disaster site would have
reached the public. With each press leak about disturbing
results from an EPA test, the agency responded by expanding
the amount of information on its web site.

Still, the efforts of Kupferman and the various community
leaders, even the complaints of thousands of people who
were getting sick, failed to convince most of the mass media
or the political leaders in the city to examine seriously the
government's handling of the crisis. At best, the pressure
convinced the city council and the state legislature to con-

duct a series of public hearings, though these produced virtually nothing in the way of concrete action.

What did have an enormous impact, however, were several reports from respected scientists who began openly to contradict the official version of the EPA and the city. Those scientists who have already been mentioned, Cate Jenkins of the EPA, Piotr Chmielinski of H.P. Environmental, and Tom Cahill of the University of California at Davis, took enormous risks in challenging the government's line. Jenkins, as an EPA employee, risked being ostracized and targeted within her own agency. And Chmielinski was immediately fired from his consulting job. Unlike these scientific heroes, several environmental experts in New York City chose to parrot the claims of the EPA and the city, and then quickly submitted applications for big government grants to conduct research on the environmental consequences of the collapse. Several such studies are now in the works.

In addition, a handful of political leaders dared to stick out their necks and question what was happening. Foremost among them were Jerrold Nadler, the veteran congressman who represents lower Manhattan, and Kathryn Freed, former City Councilwoman, who spearheaded the Ground Zero Task Force of Elected Officials. Nadler's consistent criticism of the failures of the agency and the city were instrumental in getting the EPA's ombudsman to initiate his own investigation of the federal agency. At the time, ombudsman Martin was already embroiled in a nasty battle with Whitman, who was trying to squash his tiny office by placing it under direct control of the EPA's inspector general. Martin managed to stave off the merger for several months when he convinced a federal judge to issue a temporary restraining order against the move. Meanwhile, he organized several high-profile public hearings in New York City on Ground Zero health issues that helped break the media silence. In mid-April, however, the court lifted its TRO and Whitman finally got her way. By ordering Martin to clear all further public statements with the inspector general, Whitman thus silenced her biggest critic within the agency.

Another pivotal figure in this affair was Senator Hillary Clinton. On October 26, 2001, the day my front-page column appeared in the *Daily News* revealing that the EPA had found high levels of a variety of toxins in the air and water around Ground Zero, Clinton fired off a letter to Whitman requesting more information about what the agency had found.

In the weeks that followed, Clinton and her staff pursued the issue vigorously and eventually convinced Connecticut Senator Joe Lieberman to convene a hearing of a Senate subcommittee he chairs to examine the environmental problems at Ground Zero. It was that hearing, held on February 11, 2002, and attended by an audience of more than 500 people, that finally forced Whitman to admit her agency's credibility problem. The next day she announced the establishment of a new federal task force on indoor air quality.

Seven months after the worst environmental catastrophe in urban American history, not a single top government health official has yet admitted that thousands of New Yorkers were exposed to significant health risks on September 11 and in the months that followed, or that the Giuliani administration and the federal government made enormous mistakes in sending people back to work in lower Manhattan so soon after the event, or that they failed to properly protect rescue workers during the course of the massive cleanup. As for the New York press, for the most part they obediently followed the line they were fed.

Yet in early March of 2002 a poll of New York residents conducted by the *Daily News* and New York 1, the city's all-news cable station, made a startling discovery: 70 percent of New Yorkers did not believe the EPA's assurances that the air around Ground Zero was safe. The people of New York had roundly rejected the toxic deception. Unfortunately, the cost of that deception remains to be paid.

—April 11, 2002

APPENDICES: NEW YORK CITY DEPARTMENT OF HEALTH AND EPA DOCUMENTS

HEALTH DEPARTMENT OFFERS RECOMMENDATIONS FOR INDIVIDUALS RE-OCCUPYING COMMERCIAL BUILDINGS AND RESIDENTS RE-ENTERING THEIR HOMES

(NYC Department of Health Press Release, September 17, 2001)

The New York City Department of Health announced today recommendations for building owners/managers and individuals who are being allowed to re-enter their businesses or homes for the first time since the World Trade Center disaster on September 11. The Health Department has been collaborating with numerous City, State, and Federal agencies to monitor potential public health impacts in the general vicinity of the World Trade Center blast zone.

New York City Health Commissioner Neal L. Cohen, M.D., said, "As some buildings near the World Trade Center may have sustained structural damage, experienced power loss, and/or been subject to migrating dust and debris from the blast, the Health Department working with numerous agencies has been actively monitoring the condition of buildings in and around the blast area to determine when occupants may safely resume tenancy. Property owners and managers are being instructed to assess the stability and safety of their buildings. This includes checking and, if necessary, restoring utility services."

Recommendations for Re-entry into Your Homes

In your home, first make sure that conditions are safe. Enter your home dressed in a long-sleeve shirt and pants, and with closed shoes. Upon entry:

- Check for the smell of gas. If the apartment smells of gas, leave immediately and report it to your building manager and to Con Edison.
- Check for broken glass and fixtures. Wrap any broken glass in paper and mark it "broken glass." If large pieces of glass are broken, ask your building superintendent for help.
- Run hot and cold water from each of the taps for at least two minutes, or until water runs completely clean, whichever is longer.
- Flush toilets until bowls are refilled. For air pressure systems, you may need to flush several times. If there are any problems with the toilet or plumbing systems, call a plumber — do not try to fix the problem yourself.

Recommendations for Food Left in Homes and Office Spaces

The power outage in much of lower Manhattan may have caused refrigerated and frozen food to spoil. Raw or cooked meat, poultry and seafood, milk and milk-containing products, eggs, mayonnaise and creamy dress-

* The Main Office of the NYC DOH has been temporarily relocated to 455 1st Avenue.

ings, and cooked foods should be thrown out if power was out for two or more hours. Frozen foods that have thawed should be thrown away. Do not re-freeze thawed food. Throw away any food that may have been contaminated with dust, except for food in cans, jars, or containers with tight-fitting lids. Wash dust-covered cans and jars with water and wipe clean. When it comes to food left in your building, *if in doubt, throw it out.*

Recommendations for Cleaning Homes and Office Space

The best way to remove dust is to use a wet rag or wet mop. Sweeping with a dry broom is not recommended because it can make dust airborne again. Where dust is thick, directly wet the dust with water, and remove it in layers with wet rags and mops. Dirty rags can be rinsed under running water (try not to leave dust in the sink to dry). Used rags and mops should be put in plastic bags while they are still wet and bags should be sealed and discarded. Cloth rags should be washed separately from other laundry. Wash heavily-soiled or dusty clothing or linens twice. Remove lint from washing machines and filters in the dryers with each laundry load. Rags should not be allowed to dry out before bagging and disposal or washing.

Because the dust particles are so small, standard vacuuming is not an efficient way to remove the dust and may put dust back into the air where it can be inhaled. HEPA (high efficiency particulate) filtration vacuums capable of trapping very fine particles can be used. If a HEPA vacuum is not available, either HEPA bags or dust allergen bags should be used with your regular vacuum. Carpets and upholstery can be shampooed and then vacuumed.

If your apartment is very dusty, you should wash or HEPA vacuum your curtains. If curtains need to be taken down, take them down slowly to prevent making dust in the air. To clean plants, rinse leaves with water. Pets may be washed with running water from a hose or faucet; their paws should be wiped to avoid tracking dust inside the home.

Additional recommendations include:

- Avoid sweeping or other outdoor maintenance
- Keep outdoor dust from entering the home
- Keep windows closed
- Set the air conditioner to re-circulate air (closed vents), and clean or change the filter frequently
- Remove shoes before entering the home for several days (once you first make sure there is no broken glass)
- Air purifiers may help reduce indoor dust levels. Air purifiers are only useful for removing dust from the air. They will not remove dust already deposited on floors, shelves, upholstery, or rugs. Keep windows closed when using an air purifier.

New York City/World Trade Center Sampling Activities
Street Runoff Asbestos Results

(EPA Daily Monitoring Report, September 18, 2001)
Preliminary Data

EPA Personnel:	Dennis McChesney Stephen Hale
Sampling platform:	EPA Whaler/NYCFD Fire boat (tied)
Sampling Date:	09/14/01
Sampling Time:	1430–1500
Location:	Foot of Rector St. at Hudson River. Samples were collected from approx 1-foot diameter direct runoff pipe
Sample Matrix:	Water
Analytes (Laboratory):	Metals (Region 2) PCBs/PBBs (Axys) PAHs (Axys) Dioxins/Furans (Axys) Asbestos (NYSDOH)

Sampling: Discrete grab samples of washdown water discharging to the Hudson River were collected as it flowed out of the pipe, before entering the river. Sanitation workers were hosing the street in the vicinity.

Preliminary Asbestos Results reported by Dr. James Weber, DOH—Wadsworth

Chrysotile Asbestos:	0.61% 9.6 billion fibers/L 0.045 g/L
Amphibole Asbestos:	ND
Particulates:	7.4 g/L
Blank Sample Bottle:	ND

>1% is regulated in building material, MCL 7 MFL, 10–4 cancer risk 700 MFL

New York City/World Trade Center Sampling Activities Street Runoff Results

(EPA Daily Monitoring Report, September 20, 2001)
Preliminary Data

Preliminary Results Summary:
All analyzed dioxins and furans were detected. The Toxic Equivalency (TEQ) for the sample was 122 pg/L, which is high. Toxic PCBs congeners were also detected at very high concentrations, with a TEQ of 151. PAHs were detected at concentrations similar to those in "average" NYC CSO effluent. Metals and asbestos were detected at high concentrations. The low flow and rapid dilution of the sampled discharge suggests that the water quality impact is minimal. Additional runoff (if observed) and ambient water sampling is being conducted by EPA during the storms on 09/19/01. Additional sampling of street dust is recommended.

EPA Personnel:	Dennis McChesney
	Stephen Hale
Sampling Date:	EPA Whaler/NYCFD Fire boat (tied)
Location:	Foot of Rector at Hudson River. Samples were collected from approximately one (1) foot diameter direct runoff pipe
Sample Matrix:	Water
Analytes (Laboratory):	Metals (Region 2)
	PCBs/PBFs (Axys)
	PAHs (Axys)
	Dioxins/Furans (Axys)
	Asbestos (NYSDOH)

Sampling: Discrete grab samples of washdown water discharging to the Hudson River were collected as it flowed out of the pipe, before entering the river. Sanitation workers were hosing the street in the vicinity. Samples were generally turbid and contained high solid concentrations. Dilution was observed to occur (non-observable milky color) within 25 feet of the discharge.

Dioxins/Furans: All of the target dioxins and furans were detected in the runoff sample. The highest concentration detected was for the chlorinated octachlrodibenzo dioxin (OCDD) at 5.5 pg/L. Evaluation using NATO toxic equivalency factors (TEF) show toxicity dominated by the dibenzo furan 2,3,4,7,8-PeCDF with a TEQ over 50 pg/L. The total TEQ for the sample is 122 pg/L (blank corrected). In previous harbor work performed by NYSDEC for the CARP, the highest observed dioxin TEQ was 22 pg/L. That occurred in the mid-tidal Passaic.

This figure shows the contributions of the 17 individual congeners to the total TEQ.

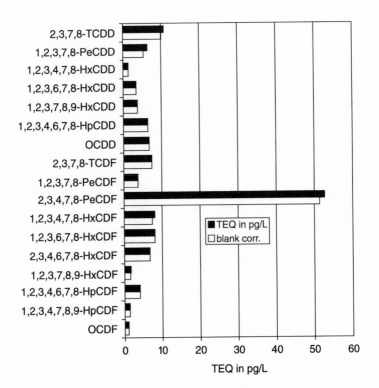

PCBs: Numerous PCB congeners including co-planer (dioxin-like) PCBs were detected at high concentrations. The Toxic Equivalency (TEQ) calculated for the toxic is PCBs 151 pg/L. In previous harbor work performed by NYSDEC for the CARP, the highest observed PCB TEQ was 0.002 pg/L.

PAHs: All PAHs were detected at high concentrations.
nanomoles/L

WTC Street Runoff   "Typical" CSO Effluent   Typical STP Effluent   Landfill Leachate

New York City/World Trade Center Sampling Activities
Street Runoff Results

(Excerpt from EPA Daily Monitoring Report, November 1, 2001)

Bulk/Dust Samples
- NYC / ER (September 11) — Pesticides/PCBs
  o Recent analysis of 4 dust samples originally collected from streets on September 11.
  o All samples below the EPA residential cleanup guideline of 1 ppm for PCBs.
- NYC / ER (September 11) — Semi-volatile organic compounds (base neutral acid extractable)
  o Recent analysis of 4 dust samples originally collected from streets on September 11.
  o All samples were below the EPA Removal Action guideline levels (based on a 30-year exposure) represented as Toxic Equivalency Factors (TEFs) for benzo(a)pyrene.

New York City/World Trade Center Sampling Activities Street Runoff Results

(Excerpt from EPA Daily Monitoring Report, December 14, 2001)

Bulk/Dust Samples
- NYC / ER (September 11)—PCB congeners
  - o Congener analysis of 4 dust samples originally collected from streets near WTC on September 11.
  - o All samples were below the EPA residential cleanup guidance of 1 ppm for PCBs.
  - o *Note*: A pesticide/PCB scan previously conducted for these four samples and presented in the November 1, 2001, Sampling Situation Report incorrectly identified all levels as being below 1 ppm. Two of the samples were actually estimated to be above 1 ppm for total PCBs. The highest total PCBs result of these two samples was estimated at 1.54 ppm.

Index